**Advanced Courses in Mathematics
CRM Barcelona**

Centre de Recerca Matemàtica

Managing Editor:
Carles Casacuberta

More information about this series at http://www.springer.com/series/5038

Giovanna Citti • Loukas Grafakos • Carlos Pérez
Alessandro Sarti • Xiao Zhong

Harmonic and Geometric Analysis

Editor for this volume:
Joan Mateu (Universitat Autònoma de Barcelona)

 Birkhäuser

Giovanna Citti
Dipartimento di Matematica
Università di Bologna
Bologna, Italy

Loukas Grafakos
Department of Mathematics
University of Missouri
Columbia, MO, USA

Carlos Pérez
Departamento de Análisis Matemático
Universidad de Sevilla
Sevilla, Spain

Alessandro Sarti
Dipartimento di Elettronica, Informatica
  e Sistemistica
Università di Bologna
Bologna, Italy

Xiao Zhong
Department of Mathematics and Statistics
University of Jyväskylä
Jyväskylä, Finland

ISSN 2297-0304   ISSN 2297-0312   (electronic)
Advanced Courses in Mathematics - CRM Barcelona
ISBN 978-3-0348-0407-3   ISBN 978-3-0348-0408-0   (eBook)
DOI 10.1007/978-3-0348-0408-0

Library of Congress Control Number: 2015938758

Mathematics Subject Classification (2010): Primary: 42B20, 42B37; Secondary: 35R03

Springer Basel Heidelberg New York Dordrecht London

Printed on acid-free paper

Springer Basel AG is part of Springer Science+Business Media (www.birkhauser-science.com)

Contents

2   Multilinear Calderón–Zygmund Singular Integrals

Loukas Grafakos

3   Singular Integrals and Weights

Carlos Pérez

4   De Giorgi–Nash–Moser Theory

Xiao Zhong

Foreword

From February to July 2009, the CRM organised a research programme entitled *Harmonic Analysis, Geometric Measure Theory and Quasiconformal Mappings*. As part of the programme, several advanced courses were delivered by leading experts in the respective fields. This volume contains expositions of the material covered during some of the courses.

As the title suggests, the aim of the courses was to deal with selected topics in harmonic analysis and partial differential equations. The first chapter of these notes, written by G. Citti and A. Sarti, explains how the Heisenberg group can be used to model human vision; regularity results for some PDE naturally arising are thoroughly discussed. In the second chapter, L. Grafakos presents some aspects of the basic theory of multilinear harmonic analysis and explains recent developments of the subject. In the third chapter, C. Pérez offers a short introduction to some aspects of the theory of singular integral operators, including a number of new results on sharp weighted bounds for Calderón–Zygmund type operators. In the last chapter, X. Zhong presents the De Giorgi–Nash–Moser theory on regularity of second-order, linear, elliptic equations in divergence form.

Thanks are due to the Centre de Recerca Matemàtica for organising and sponsoring the research programme, and to the Centre's administrative staff for smoothly working out innumerable details. Finally, we are grateful to all the participants for their interest in the event and for their positive response.

Joan Mateu

Chapter 1

Models of the Visual Cortex in Lie Groups

Giovanna Citti and Alessandro Sarti

1.1 Introduction

The most classical and exhaustive theory which states and studies the phenomenological laws of visual reconstruction is Gestalt theory [73, 74]. It formalizes visual perceptual phenomena in terms of geometric concepts, such as good continuation, orientation, or vicinity. Consequently, phenomenological models of vision have been expressed in terms of geometrical instruments and minima of calculus of variations [5, 51, 96]. On the other hand, the recent progress of medical imaging and integrative neuroscience allows to study neurological structures related to perception of space and motion. The first results that used instruments of differential geometry to model the cortex and justify macroscopical visual phenomena in terms of the microscopical behavior of the cortex were due to Hoffmann [70], Koenderink [77], and Petitot–Tondut [100, 102]. More recently, in [37] and [109], the visual cortex was modeled as a Lie group with a sub-Riemannian metric. Other models in Lie groups are proposed in [12, 22, 39, 49, 62, 63, 112, 117, 118]. We refer to these papers for a complete description of this type of problems.

The variables coded in the cortex are orientation of boundaries, velocity or direction of movement, which are related with differential constraints to the other variables of the space. These differential constraints define a sub-bundle of the tangent space at every point of the space, which, together with a Riemannian metric on it, defines a sub-Riemannian structure. We will as well assume that the basis $X_1, \ldots, X_m$ of the sub-bundle satisfies the celebrated Hörmander condition [71], which ensures that the Lie algebra generated by the vector fields (X_i) has maximum rank at every point. Under this condition, the Chow Theorem is satisfied and every couple of points can be joined by an integral curve of the vector fields (X_i). Hence a control distance can be defined, called Carnot–Carathéodory metric.

The measure of the balls of the metric has been studied by Nagel, Stein and Wainger [93]. In this context, the vector fields (X_i) play a role analogous to the derivatives in the Euclidean setting. All the differential calculus can be restated in this way. The study of PDE represented as sums of squares of vector fields goes back to the papers of Hörmander [71], Rothschild and Stein [107], and Folland [56], who proved the hypo-ellipticity of the subelliptic Laplacians and studied properties of their fundamental solution.

The notion of regular surface in this setting is a much more recent concept: it has been introduced by Franchi, Serapioni and Serra Cassano in [59]. Surprisingly, the surface cannot be defined as the image of a subset of Euclidean space, as initially conjectured by Ambrosio and Kirchheim [1]. The implicit function theorem has been proved in this setting by Ambrosio, Serra Cassano and Vittone [3], and by Citti and Manfredini [35]. It gives rise to non-linear vector fields. This is why the mean curvature of a regular surface is a second-order differential equation expressed in terms of non-linear vector fields. This expression has been introduced with different instruments by different authors; see [3, 30, 41, 98, 106]. However, the problem of regularity of minimal surfaces is largely open. Preliminary results have been proved in [24, 25].

Here we give a presentation of the model of Citti–Sarti, together with the instruments of sub-Riemannian differential geometry necessary for its description, and the results which support the model. The main goal is to underline why sub-Riemannian geometry is a natural instrument for the description of the visual cortex.

In Sections 1.2 and 1.3 we describe the problem of perceptual completion and give a short description of the functional architecture of the visual cortex. In Section 1.4 we describe the functional geometry of the visual cortex as a sub-Riemannian structure and give the principal definition and properties of a sub-Riemannian space. In Section 1.5 we give an introduction of differential calculus in Lie groups, define a uniformly sub-Riemannian operator, and its time-dependent counterpart. Then we show that these operators can model the propagation of the visual signal in the cortex. In Section 1.6 we study the regular surfaces of the structure and prove that the neural mechanism of non-maxima suppression generates regular surfaces in the cortical space. Finally, in Section 1.7 we prove that the two mechanisms of propagation of the visual signal and non-maxima suppression generate a diffusion-driven motion by curvature. The perceptual completion is then obtained through a minimal surface. Hence we will study its regularity and foliation properties.

This review has been prepared in 2009 and it is the text of a course given in Barcelona in the same year. In the meanwhile other results have been obtained in this neuromathematical and sub-Riemannian setting. Let us quote the cortical models developed in [7, 13, 15, 78, 110] and the monograph [38]. From the point of view of image processing see for example [19, 47, 48, 50, 55, 69, 97]. For properties of minimal surfaces we refer to [28, 29, 31, 43, 64, 75, 90, 91, 104, 106, 114].

1.2 Perceptual completion phenomena

Gaetano Kanizsa provided in [73, 74] a taxonomy of perceptual completion phenomena and outlined that they are interesting tests to understand how the visual system interpolates existing information and builds the perceived units.

He discriminated between modal completion and amodal completion. In the first one, the interpolated parts of the image are perceived with the full modality of the vision and are phenomenally undistinguishable from real stimuli (this happens for example in the formation of illusory contours and surfaces). In amodal presence, the configuration is perceived without any sensorial counterpart. Amodal completion is evoked every time one reconstructs the shape of a partially occluded object. Thus it is at the base of the most primitive perceptual configuration, that is, the segmentation of figure and ground. Mathematical models of perceptual completion take into account main phenomenological properties as described by the psychology of Gestalt.

1.2.1 Gestalt rules and association fields

The history of studies on contour integration is a long one, stretching back to the Gestalt psychologists who formulated rules for perceptually significant image structure, including contour continuity: the Gestalt law of good continuation. Field, Hayes and Hess [54] developed a new approach to psychophysically investigating how the visual system codes contour continuity by using contours of varying curvature made up of spatial frequency narrowband elements. The contour stimulus is shown in Fig. 1.1. Within a field of evenly spaced, randomly oriented, Gabor elements, a subset of the elements is aligned in orientation and position along a notional contour (Fig. 1.1 A). This stimulus is paired with a similar stimulus (Fig. 1.1 B), where all of the elements (called background elements) are unaligned. The observer was asked to recognize structures and alignments in the stimulus, and to discriminate the two stimuli. From a simple informational point of view, Figs. 1.1 A and B are equivalent, so a difference in their detectability reflects the ability of human observers to detect the contour and constraints imposed by the visual system. In particular, it is interesting to note that contours composed of elements whose local orientation was orthogonal to the contour are far less detectable.

Another finding of this study was the human ability to detect increasingly curved contours. A good performance for contour detection was possible even in presence of curvature of the contour, suggesting that the output of cells with similar, but not necessarily equal orientation preference are being integrated. Fig. 1.1 C shows another stimulus manipulation that reinforces the notion that the task of contour integration reflects the action of a network rather than that of single neurons interaction. Here the polarity of every other Gabor element is flipped. The contour (and background) is now composed of Gabor elements alternating in their contrast polarity. The visibility of the contour in Figs. 1.1 A and C is similar.

Psychophysical measurement shows that although there is a small decrement in performance in the alternating polarity condition, curved contours are still readily detectable when composed of elements of alternating polarity.

This model of cellular interaction and contour completion has been summarized by Field, Hayes and Hess in terms of an association field which is depicted in Fig. 1.2. The stimulus in the central position can be joined with other stimula tangent to the lines in the figure, but cannot be joined with stimula with a different direction.

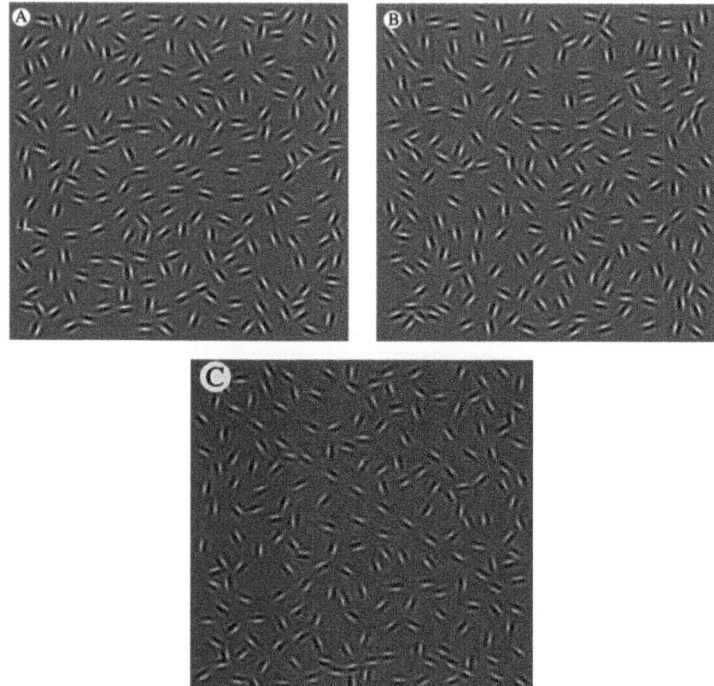

Figure 1.1: The experiment of Field, Hayes and Hess. (Reproduced with permission from [67].)

1.2.2   The phenomenological model of elastica

Since subjective boundaries could be linear or curvilinear, their reconstruction is classically performed by minimizing the elastica functional

$$\int_\gamma (1 + k^2)\, ds, \tag{1.1}$$

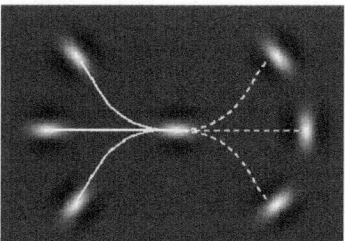

Figure 1.2: Association field. (Reproduced with permission from [67].)

where the integral is computed on the missed boundary, and k is its curvature (see [92]). The minimum of the elastica functional is taken on all curves with fixed endpoints and with fixed directions at the endpoints. It appears that continuation of object boundaries plays a central role in the disocclusion process. This continuation is performed between T-junctions, which are points where image edges intersect orthogonally as illustrated in Fig. 1.3.

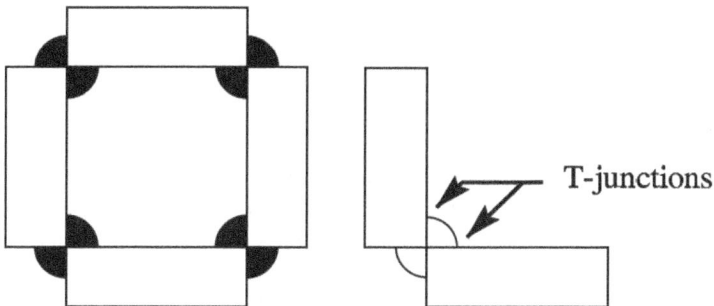

Figure 1.3: An example of T-junction.

In [96], Mumford, Nitzberg and Shiota deduced from the amodal completion principles a method for detecting and recovering occluded objects in a still image within the framework of a segmentation and depth computing algorithm.

Approximation in the sense of Γ-convergence by elliptic functionals was proposed by De Giorgi in [46] (the conjecture is still open). Bellettini and Paolini [11] proposed and proved a new approximation of Modica–Mortola type. They also proved that the functional (1.1) does not allow non-regular completion, which in fact can occur, and proposed to modify the functional with a new functional

$$\int_\gamma (1 + \phi(k^2)) \, ds.$$

When ϕ has linear growth at the origin and behaves as a square root at infinity, completion with kinks is allowed.

The extension of the elastica functional to the level set of the image I has been applied in problems of inpainting (that can be considered a particular case of modal completion) by [2, 83]:

$$\int_{\Omega} |\nabla I| \left(1 + \left| \operatorname{div} \left(\frac{\nabla I}{|\nabla I|} \right) \right|^2 \right) dx, \quad x \in \Omega \subset \mathbb{R}^2,$$

where the integral is extended to the domain of the image. In this way each level line of the image is completed either linearly or curvilinearly as elastica curve.

In order to make occluded and occluding objects present at the same time in the image, in [96] (and then in [10, 51]) a third dimension is introduced, and the objects present in the image are represented as a stack of sets, ordered by depth. In [113], the third added dimension is represented by time, and the algorithm first detects occluding objects, then occluded ones. In [5], the associated evolution equation was split into two equations, each one of first order and depending on two different variables: the image I and the direction of its gradient $\nu = \nabla I/|\nabla I|$.

1.3   Functional architecture of the visual cortex

From the neurophysiological point of view, the acquisition of the visual system is performed in the retina, that, after a preprocessing, projects the information to the lateral geniculate nucleus and to the primary visual cortex, in which signal is deeply processed. In particular, the primary visual cortex V1 processes the orientation of contours by means of the so-called simple cells and other features of the visual signal by means of complex cells (stereoscopic vision, estimation of motion direction, detection of angles). Every cell is characterized by its receptive field, that is, the domain of the retinal plane to which the cell is connected with neural synapses of the retinal-geniculate-cortical path. When the domain is stimulated by a visual signal, the cell responds generating spikes.

Classically, a receptive profile is subdivided in "on" and "off" areas. The area is considered "on" if the cell spikes responding to a positive signal, and "off" if it spikes responding to a negative signal. The receptive profile is mathematically described by a function Ψ_0 defined on the retinal plane. This function models the neural output of the cell in response to a punctual stimulus in the 2-dimensional point x. Simple cells have directional receptive profiles as shown in Fig. 1.4 and they are sensitive to the boundaries of images.

To understand the processing of the image operated by these cells, it is necessary to consider the functional structures of the primary visual cortex: the retinotopic organization, the hypercolumnar structure with intracortical circuitry and the connectivity structure between hypercolumns.

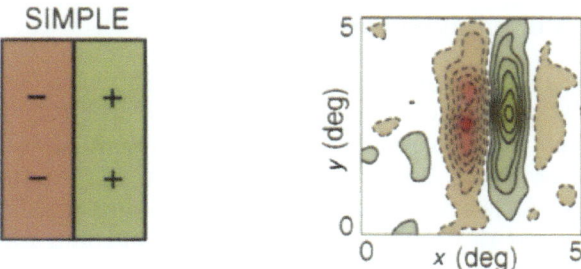

Figure 1.4: Receptive profiles. (Reproduced with permission from [45].)

1.3.1 The retinotopic structure

The retinotopic structure is a mapping between the retina and the primary visual cortices that preserves the retinal topology and is mathematically described by a logarithmic conformal mapping. From the image processing point of view, the retinotopic mapping introduces a simple deformation of the stimulus image that will be neglected in the present study.

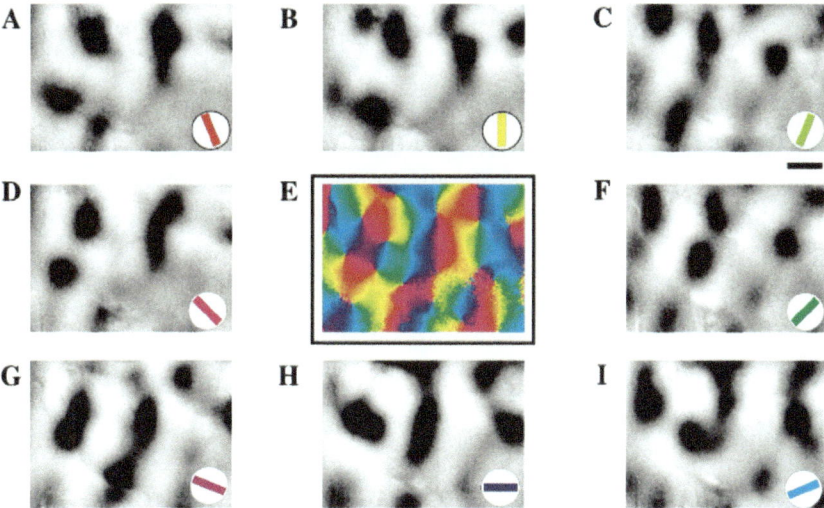

Figure 1.5: Representation of Bosking. Within a hypercolumn the cells sensible to different orientations are represented in different colours. (Reproduced with permission from [40].)

1.3.2   The hypercolumnar structure

The hypercolumnar structure organizes the cortical cells in columns corresponding to parameters like orientation, ocular dominance, color, etc. For the simple cells (sensitive to orientation), columnar structure means that to every retinal position it is associated a set of cells (hypercolumn) sensitive to all the possible orientations. The visual cortex is indeed two-dimensional and then the third dimension collapses onto the plane giving rise to the fascinating pinwheel configuration observed by William Bosking et al. with optical imaging techniques. In Figs. 1.5, the orientation preference of cells is coded by colors and every hypercolumn is represented by a pinwheel.

1.3.3   The neural circuitry

The *intracortical circuitry* is able to select the orientation of maximum output of the hypercolumn in response to a visual stimulus and to suppress all the others. The mechanism able to produce this selection is called non-maxima suppression or orientation selection, and its deep functioning is still controversial, even if many models have been proposed (see [85, 94, 103]).

   The *connectivity structure*, also called horizontal or cortico-cortical connectivity is the structure of the visual cortex that ensures connectivity between hypercolumns. The horizontal connections connect cells with the same orientation belonging to different hypercolumns. Historically, correlation techniques have been used to estimate the relation between connectivity and preferred orientation of cells [116]. Only recently techniques of optical imaging associated to tracers allowed a large-scale observation of neural signal propagation via cortico-cortical connectivity. These tests have shown that the propagation is highly anisotropic and almost collinear to the preferred orientation of the cell; see the study of Bosking et al. [20, Fig. 4]. It is already confirmed that this connectivity allows the integration process, that is at the base of the formation of regular and illusory contours and of subjective surfaces [102]. Obviously the functional architecture of the visual cortex is much richer of the schemata that we have delineated – just think of the high percentage of feedback connectivity from superior cortical areas – but for now we will try to propose a model of low-level vision, aiming to mathematically model correctly the functional structures that we have described and to be able to show that these are at the base of perceptual completion of contours.

1.4   The visual cortex modeled as a Lie group

1.4.1   A first model in the Heisenberg group

Petitot and Tondut proposed in [102] a new approach to the problem, which is particularly interesting because the perceptual completion problem is considered as a problem of naturalizing phenomenological models on the basis of biological and neurophysiological evidence. Let us recall here their model.

Retinotopic and (hyper)columnar structure   The main structures of the cortex, retinotopic and (hyper)columnar, can be modeled as follows.

- Retinotopy means that there exist mappings from the retina to the cortical layers which preserve retinal topography. If we identify the retinal structure with a plane R (the retina) and denote by M the cortical layer, the retinotopy is then described by a map $q \colon R \to M$, which is an isomorphism. Hence we will identify the two planes and call M both of them.

- The columnar and hypercolumnar structure organizes the cells of V1 in columns corresponding to orientation. Due to their RP, they detect preferred orientations, that is, points (x, u) where x denotes a 2-dimensional (retinal) position and u denotes the direction of a boundary of an image mapped on the retina at the point x.

  The hypercolumnar organization means essentially that to each position x of the retina there exists a full fibre of possible orientations u at x.

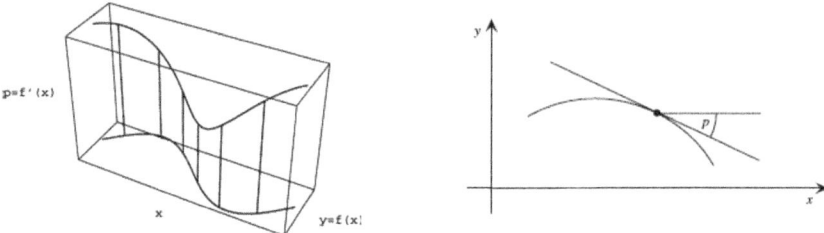

Figure 1.6: Curves lifted in the cortical contact structure.

Contour detection and lifting   Formally, at a retinal point $x = (x_1, x_2)$ we consider edges of images as regular curves of the form

$$x_2 = f(x_1).$$

The orientation at the point x is then $u = f'(x_1)$. The tangent vector to the considered edge at the point x has the expression

$$X_u = \partial_1 + u(x_1)\partial_2.$$

In presence of the visual stimulus, all the hypercolumn over the point x is activated, and the simple cell sensible to the direction u has the maximal response. The retinal point x is lifted to the cortical point (x, u), and the whole curve is then lifted to the curve

$$(x_1, f(x_1), u(x_1))$$

in a 3-dimensional space $\mathbb{R}^3$ endowed with the constraint $f' = u$. Formally, this is a constraint on the tangent space $T\mathbb{R}^3$ at every point. We can define a 1-form

$$\omega = dx_2 - u\,dx_1,$$

and note that all the lifted curves lie in the kernel of ω. This formal constraint can be expressed by saying that we consider a subset of the tangent plane (kernel of the 1-form ω)

$$HT = \{\alpha X_1 + \beta X_2\},$$

where

$$X_1 = \partial_1 + u\partial_2, \quad X_2 = \partial_u. \tag{1.2}$$

The lifted curves have to be integral curves of the vector fields X_1 and X_2.

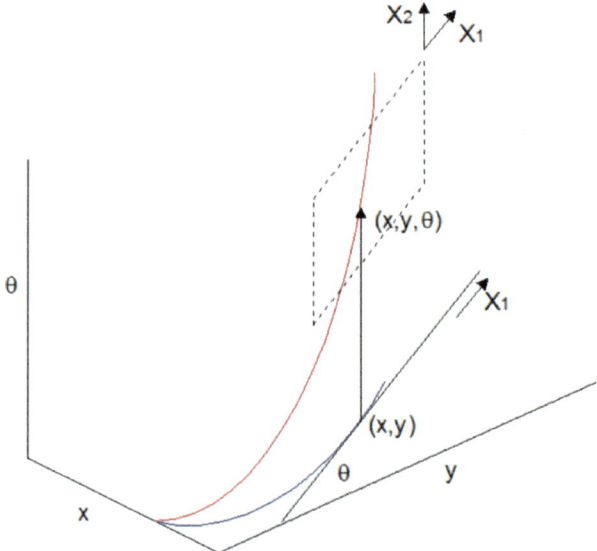

Figure 1.7: A lifted surface, foliated in lifted curves.

1.4.2   A sub-Riemannian model in the rototranslation group

The previous model can describe only images with equi-oriented boundaries. This can be easily overcome in the $E(2)$-group of motions of the plane. In [37] we recognize the previously described structure as a sub-Riemannian structure. Besides we will focus on level lines representation instead of edge detection. Indeed, if $I(x)$ is a gray level image, the family of level lines is a complete representation of I, from which I can be reconstructed. This model is compatible with the functionality of the simple cells and their orientation sensitivity.

Lifting in $EO(2)$ – a purely perceptual description   We now consider a real stimulus, represented as an image I. We can assume that cells over each point x can code the direction of the level lines of I, without a preferred direction. Hence the eingrafted variable in the hypercolumn will be an angle, and we will assume that the cell which gives the maximal response is sensible to the direction $\theta(x) = -\arctan(I_1/I_2)$, $\theta \in [0, \pi]$. This means that the vector field

$$X_\theta = \cos(\theta(x))\partial_1 + \sin(\theta(x))\partial_2 \tag{1.3}$$

is tangent to the level lines of I at the point x. As before, this process associates to each retinal point x the three-dimensional cortical point $(x, \theta) \in \mathbb{R}^2 \times S^1$. Since the process is repeated at each point, each level line is lifted to a new curve in three-dimensional space.

The tangent vector to the lifted curve can be represented as a linear combination of the vectors

$$X_1 = \cos(\theta)\partial_1 + \sin(\theta)\partial_2, \quad X_2 = \partial_\theta. \tag{1.4}$$

The set of vectors

$$a_1 X_1 + a_2 X_2$$

defines a plane and every lifted curve is tangent to a vector of the plane.

The lifting process – a neurophisiological description   Neural evidence supports this model of the cortex. When a visual stimulus of intensity $I(x)$ activates the retinal layer of photoreceptors $M \subset \mathbb{R}^2$, the cells centered at every point x of M process in parallel the retinal stimulus with their receptive profile, which is a function defined on M.

Each RP depends upon a preferred direction θ and it has been observed experimentally that the set of simple cell RPs is obtained via translations and rotations from a unique profile of Gabor type (see for example Jones–Palmer [72], Daugman [44], Marcelja [82]). This means that there exists a mother profile Ψ_0 from which all the observed profiles can be deduced by rigid transformation.

A good formula for Ψ_0 seems to be (see Fig. 1.9 and compare with Fig. 1.4)

$$\Psi_0(x) = \partial_2 e^{-|x|^2}.$$

Therefore, by rotation, all the observed profiles over the same point can be modeled as

$$\Psi_\theta(x) = \Psi_0\Big(x_1 \cos\theta + x_2 \sin\theta, -x_2 \sin\theta + x_2 \cos\theta\Big).$$

In the rotation of an angle θ, the derivative ∂_2 becomes

$$X_3 = -\sin(\theta)\partial_1 + \cos(\theta)\partial_2.$$

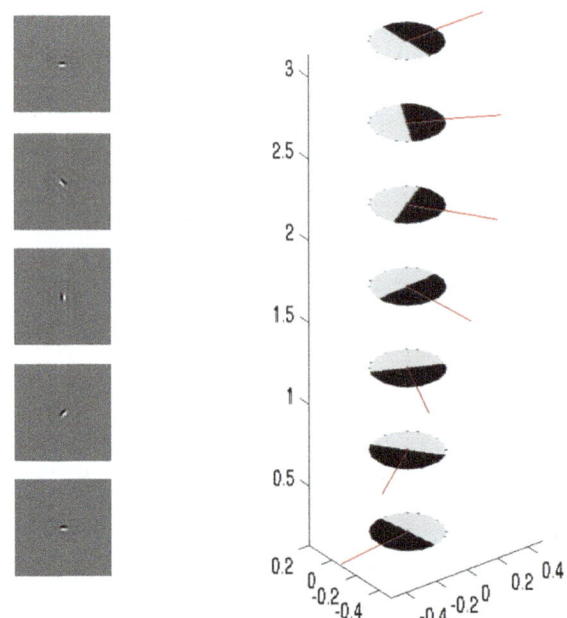

Figure 1.8: Odd part of Gabor filters with different orientations (left) and schemata of odd simple cells arranged in a hypercolumn of orientations.

Hence,

$$\Psi_\theta(x) = X_3 \, e^{-|x|^2}.$$

On the other hand, the expression of filters on different points is obtained by translation:

$$\Psi_{x,\theta}(\widetilde{x}) = \Psi_\theta(x - \widetilde{x}).$$

With this notation, the filtering can be described as convolution with the image I and generates a function

$$O(x,\theta) = \int \Psi_{x,\theta}(\widetilde{x}) I(\widetilde{x}) \, d\widetilde{x} = -X_3 \exp(-|x|^2) * I = -X_3(\theta) I_s, \qquad (1.5)$$

where we have denoted by I_s the convolution of I with a smoothing kernel,

$$I_s = I * \exp(-|x|^2).$$

This function O is the output of the cells and measures their activity. Note that $O(x,\theta)$ depends on the orientation θ. Due to the expression of the Gabor filter,

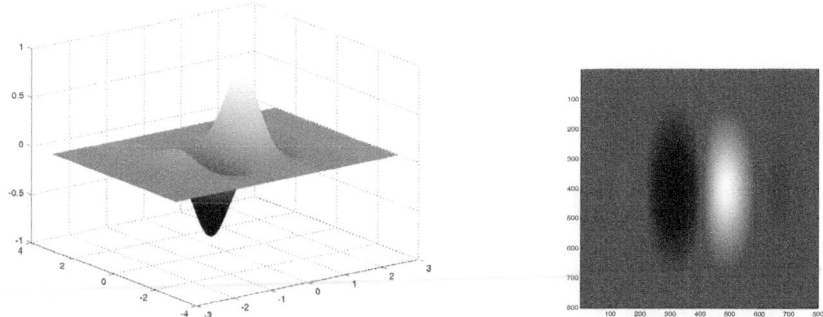

Figure 1.9: The shape of the Gabor filter and a schematic representation of it; compare with the in vivo registration, Fig. 1.4.

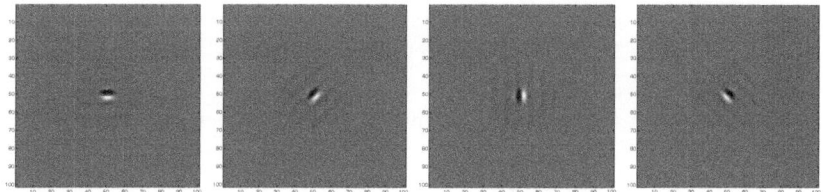

Figure 1.10: Odd part of Gabor filters with different orientations: $\theta = 0$, $\theta = \pi/4$, $\theta = \pi/2$, $\theta = 3/2\pi$.

the function O exponentially decays from its maxima. Hence, for θ fixed, it selects a neighborhood of the points where the component of the gradient in the direction $(-\sin\theta, \cos\theta)$ is sufficiently big (see Fig. 1.11).

The convolution mechanism (1.5) is insufficient to explain the strong orientation tuning exhibited by most simple cells. For these reasons, the classic feedforward mechanism must be integrated with additional mechanisms, in order to provide the sharp tuning experimentally observed. The basic mechanism is controversial and in the past years several models have been presented to explain the emergence of orientation selectivity in the primary visual cortex: "push-pull" models [85, 103], "emergent" models [94], and "recurrent" models [84], only to cite a few. Nevertheless, it is evident that the intracortical circuitry is able to filter out all the spurious directions and to strictly keep the direction of maximum response of the simple cells.

We will then define

$$O(x, \bar{\theta}) = \max_{\theta} O(x, \theta).$$

This maximality condition can be mathematically expressed by requiring that the derivative of O with respect to the variable θ vanishes at the point $(x, \bar{\theta})$:

$$\partial_\theta O(x, \bar{\theta}) = 0.$$

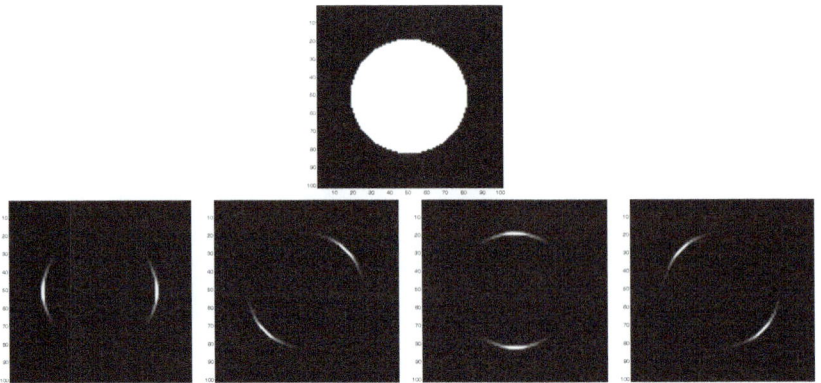

Figure 1.11: The original image showing a white disk (upper) and a sequence of convolutions with differently oriented Gabor filters.

At the maximum point $\bar{\theta}$, the derivative with respect to θ vanishes, and we have

$$0 = \frac{\partial}{\partial\theta}O(x,\bar{\theta}) = \frac{\partial}{\partial\theta}X_3(\bar{\theta})I = -X_1(\bar{\theta})I = -\langle X_1(\bar{\theta}), \nabla I\rangle.$$

As a direct consequence, we can deduce that the lifted curves are tangent to the plane generated by the vectors X_1 and X_2.

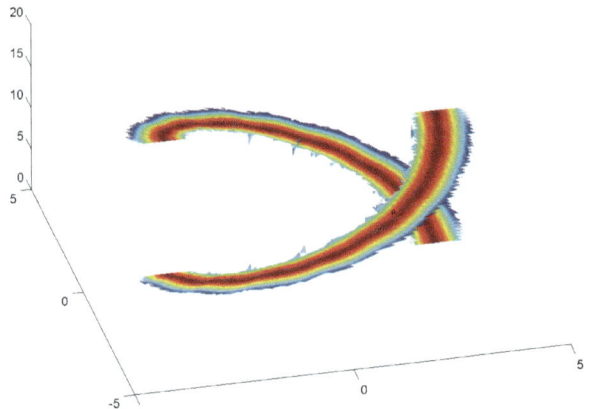

Figure 1.12: The resulting surface after non-maxima suppression, called lifted surface.

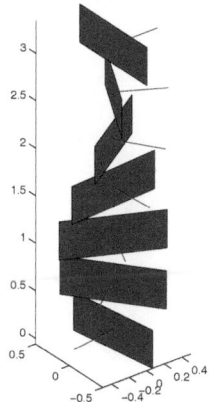

Figure 1.13: Contact planes at every point and the orthogonal vector $X_3 I$.

1.4.3 Hörmander vector fields and sub-Riemannian structures

In the standard Euclidean setting, the tangent space to $\mathbb{R}^n$ has dimension n at every point. In the geometric setting arising from the model of the cortex, the dimension of the space is 3, but we have selected at every point a 2-dimensional subspace of the tangent space, and verified that all admissible curves are tangent to this subspace at every point. We will see that these are examples of sub-Riemannian structures.

In general, we will denote by ξ the points in $\mathbb{R}^n$ and choose m first-order smooth differential operators

$$X_j = \sum_{k=1}^{n} a_{jk} \partial_k, \qquad j = 1, \ldots, m$$

in $\mathbb{R}^n$ with $m < n$ and a_{jk} of class C^∞. We call *horizontal tangent space* at the point $\xi \in \mathbb{R}^n$ the vector space $HH_{|\xi}$ spanned by these vector fields at the point ξ. The distribution of planes defined in this way is called *horizontal tangent bundle* and it is a subbundle of the tangent one. A differential operator X is called *horizontal* if it belongs to the horizontal bundle HH.

Definition 1.1. We call *horizontal norm* and *horizontal scalar product*, and denote them respectively by $\langle \cdot, \cdot \rangle_H$ and $| \cdot |_H$, the scalar product and the norm defined on the horizontal bundle which makes the basis $X_1, \ldots, X_m$ an orthonormal basis.

The horizontal tangent bundle is naturally endowed with a structure of algebra through the bracket.

Definition 1.2. If X, Y are first-order regular differential operators, their *commutator* (or *bracket*) is defined as

$$[X, Y] = XY - YX,$$

and it is also a first-order differential operator. We call *Lie algebra generated by* $X_1, \ldots, X_m$, and denote it by

$$\mathcal{L}(X_1, \ldots, X_m),$$

the linear span of the operators $X_1, \ldots, X_m$ and their commutators of any order.

We say that the vectors $X_1, \ldots, X_m$ have degree 1 and $[X_i, X_j]$ have degree 2, and define a degree in an analogous way for higher-order commutators.

Example 1.3. In general, the degree is not unique. Indeed, if we consider the vector fields introduced in (1.4), the vector X_1 has degree 1, but it also has degree 3, since in that specific example $X_1 = -[X_2, [X_2, X_1]]$.

Hence we call *minimum degree* of $X_j \in \mathcal{L}(X_1, \ldots, X_m)$ the value

$$\deg(X_j) = \min\{i \, : \, X_j \text{ has degree } i\}.$$

Since $m < n$, in general $\mathcal{L}(X_1, \ldots, X_m)$ will not coincide with the Euclidean tangent plane. If these two spaces coincide, we say that the Hörmander condition is satisfied:

Definition 1.4. Let $\Omega \subset \mathbb{R}^n$ be an open set, and let (X_j), $j = 1, \ldots, m$ be a family of smooth vector fields defined on Ω. If the condition

$$\text{rank}(\mathcal{L}(X_1, \ldots, X_m))(\xi) = n$$

is satisfied for every $\xi \in \mathbb{R}^n$, we say that the vector fields (X_j), $j = 1, \ldots, m$ satisfy the *Hörmander rank condition*.

If this condition is satisfied at every point ξ, we can find a number s such that (X_j), $j = 1, \ldots, m$ and their commutators of degree smaller than or equal to s span the space at ξ. If s is the smallest of such natural numbers, we say that the space has *step* s at the point ξ. At every point we can select a basis $\{X_j \, : \, j = 1, \ldots, n\}$ of the space made out of commutators of the vector fields (X_j), $j = 1, \ldots, m$. In general, the choice of the basis will not be unique, but we will choose a basis such that, for every point,

$$Q = \sum_{j=1}^{n} \deg(X_j) \tag{1.6}$$

is minimal. The value of Q is called the *local homogeneous dimension* of the space. In general it is not constant, but by simplicity we will assume in the sequel that s and Q are constant in the considered open set. This assumption is always satisfied in a Lie group.

Example 1.5. The simplest example of a family of vector fields is the Euclidean one: $X_i = \partial_i$, $i = 1, \ldots, m$ in $\mathbb{R}^n$. If $m = n$, then the Hörmander condition is satisfied while it is violated if $m < n$.

Example 1.6. Let us consider the family of vector fields introduced in (1.2). In this example, the points of $\mathbb{R}^3$ are denoted by $\xi = (x_1, x_2, u)$ and

$$X_1 = \partial_1 + u\partial_2, \quad X_2 = \partial_u.$$

Since $[X_1, X_2] = -\partial_2$, the Hörmander condition is satisfied.

Example 1.7. In (1.4) we denote by $\xi = (x_1, x_2, \theta)$ a point in $\mathbb{R}^2 \times S^1$ and denote by

$$X_1 = \cos(\theta)\partial_1 + \sin(\theta)\partial_2, \quad X_2 = \partial_\theta$$

the generators of the Lie algebra. The commutator is

$$X_3 = [X_2, X_1] = -\sin(\theta)\partial_1 + \cos(\theta)\partial_2,$$

which is linearly independent of X_1 and X_2.

1.4.4 Connectivity property

If X is a smooth first-order differential operator, $X = \sum_{k=1}^n a_k \partial_k$, and I is the identity map $I(\xi) = \xi$, then it is possible to represent the vector field with the same components as the differential operator X in the form

$$XI(\xi) = (a_1, \ldots, a_n).$$

Sometimes the vector and the differential operator are identified, but in this section we will keep them distinct for the convenience of the reader.

We call *integral curve* of the vector field XI starting at ξ_0 a curve γ such that

$$\gamma' = XI(\gamma), \quad \gamma(0) = \xi_0.$$

The curve will also be denoted by

$$\gamma(t) = \exp(tX)(\xi_0).$$

If X is horizontal, we call *horizontal curves* its integral curves.

The *Carnot–Carathéodory distance* in the space is defined in terms of horizontal integral curves, in analogy with the well-known Riemannian distance. Since in the sub-Riemannian setting we allow only integral curves of horizontal vector fields, we need to ensure that it is possible to connect any couple of points ξ and ξ_0 through a horizontal integral curve.

Theorem 1.8 (Chow Theorem). *If the Hörmander condition is satisfied, then any couple of points in $\mathbb{R}^n$ can be joined with a piecewise C^1 horizontal curve.*

Let us postpone the proof until after a few examples of vector fields satisfying the connectivity condition. We will consider the same examples as before.

Example 1.9. In the Euclidean case considered in Example 1.5, if $m = n$ then the Hörmander condition is satisfied and any couple of points can be joined with an Euclidean integral curve. If $m < n$, when the Hörmander condition is violated, also the connectivity condition fails. Indeed, if we start from the origin with an integral curve of the vectors $X_i = \partial_i$, $i = 1, \dots m$, we can reach only points with the last $n - m$ coordinates identically 0.

Example 1.10. In Example 1.6, the Hörmander condition is satisfied. On the other hand, it is easy to see that we can connect any point (x, u) with the origin through a piecewise regular horizontal curve. Indeed, we can call $\widetilde{u} = x_2/x_1$, and consider the segment $[(0,0), (0,\widetilde{u})]$, which is an integral curve of X_2. Then the segment $[(0,\widetilde{u}), (x,\widetilde{u})]$ is an integral curve of X_1. Finally, the segment $[(x,\widetilde{u}), (x, u)]$ is an integral curve of X_2.

Example 1.11. We already verified that the vector fields described in Example 1.7 satisfy the Hörmander condition. On the other hand, also in this case it is possible to verify directly that any couple of points can be connected by a piecewise regular path (see Fig. 1.14).

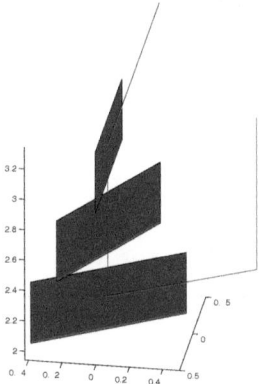

Figure 1.14: Piecewise constant integral
curves of the structure.

We follow the approach of [18] for the proof of the Chow Theorem. It is based on the following lemma:

Lemma 1.12. *If X is of class C^2, then the following estimation holds:*

$$C(t)(\xi) = e^{-tY} e^{-tX} e^{tY} e^{tX}(\xi)$$
$$= \xi + t^2(YX - XY)I(\xi) + o(t^2) = \exp(t^2[X,Y](\xi) + o(t^2))(\xi).$$

If the coefficients of the vector field X are of class C^h, we can define inductively

$$C(t, X_1, \dots, X_h)(\xi) = e^{-tX_1} C(t, -X_2, \dots, X_h) e^{tX_1} C(t, X_2, \dots, X_h)(\xi).$$

In this case we have:

$$C(t, X_1, \dots, X_h) = \exp(t^h[\dots [[X_1, X_2], \dots, X_h] + o(t^h))(\xi).$$

Proof. Let us prove the first assertion. The Taylor expansion ensures that

$$e^{tX}(\xi) = \xi + tXI(\xi) + \frac{t^2}{2} X^2 I(\xi) + o(t^2).$$

Also note that, by definition of the Lie derivative,

$$YI(\xi + tXI(\xi) + \frac{t^2}{2} X^2 I(\xi) + o(t^2))$$
$$= YI(e^{tX}(\xi) + o(t^2)) = YI(\xi) + tXYI(\xi) + o(t).$$

Hence,

$$e^{tY} e^{tX}(\xi) = e^{tY}(\xi + tXI(\xi) + \frac{t^2}{2} X^2 I(\xi) + o(t^2))$$
$$= \xi + tXI(\xi) + \frac{t^2}{2} X^2 I(\xi) + tYI(\xi + tXI(\xi) + o(t)) + \frac{t^2}{2} Y^2 I(\xi) + o(t^2)$$
$$= \xi + tXI(\xi) + \frac{t^2}{2} X^2 I(\xi) + tYI(\xi) + t^2 XYI(\xi) + \frac{t^2}{2} Y^2 I(\xi) + o(t^2)$$
$$= \xi + t(XI(\xi) + YI(\xi)) + \frac{t^2}{2} (X^2 I(\xi) + 2XYI(\xi) + Y^2 I(\xi)) + o(t^2).$$

Applying e^{-tX}, we obtain

$$e^{-tX} e^{tY} e^{tX}(\xi) = \xi + tYI(\xi) + \frac{t^2}{2} (2[X, Y]I(\xi) + Y^2 I(\xi)) + o(t^2).$$

Finally,
$$e^{-tY} e^{-tX} e^{tY} e^{tX}(\xi) = \xi + t^2[X, Y]I(\xi) + o(t^2).$$

The second assertion can be proved by induction, using the same ideas. $\square$

Proof of the connectivity property. We make the basis choice described in (1.6), and assume that $X_i = [X_{j_1}, \dots, X_{j_i}]$ for suitable indices j_i. Let us call

$$C_i(t) = C\left(t^{1/\deg(X_i)}, X_{j_1}, \dots, X_{j_i}\right).$$

By the previous lemma,

$$\frac{d}{dt} C_i(t)|_{t=0} = X_i.$$

Now, for all $e \in \mathbb{R}^n$ and $\xi \in \Omega$, we define

$$C_p(e)(\xi) = \prod_{i=1}^{n} C_i(e_i)(\xi). \tag{1.7}$$

The Jacobian determinant of C_p with respect to e is the determinant of X_i, so it is different from 0. Hence, the map $C_p(e)$ is a local homeomorphism and the connectivity property is locally proved. A connectedness argument concludes the proof. $\qquad\qquad\qquad\qquad\qquad\qquad\qquad\qquad\qquad\qquad\qquad\qquad\qquad\qquad\square$

1.4.5 Control distance

If the connectivity property is satisfied, it is possible to give the definition of distance in space. We have chosen the Euclidean metric on the contact planes, so that we can define the length of any horizontal curve γ as the value

$$\lambda(\gamma) = \int_0^1 |\gamma'(t)| \, dt.$$

Consequently, we can define a distance as

$$d(\xi, \xi_0) = \inf\{\lambda(\gamma) : \gamma \text{ is a horizontal curve connecting } \xi \text{ and } \xi_0\}. \tag{1.8}$$

Parametrizing the curve by arc length, we deduce that

$$d(\xi, \xi_0) = \inf\left\{T : \gamma' = \sum_{j=1}^{m} e_j X_j, \ \gamma(0) = \xi_0, \ \gamma(T) = \xi, \ \sqrt{\sum_{j=1}^{m} |e_j|^2} = 1\right\}$$

$$= \inf\left\{T : \gamma' = \sum_{j=1}^{m} e_j X_j, \ \gamma(0) = \xi_0, \ \gamma(T) = \xi, \ \sqrt{\sum_{j=1}^{m} |e_j|^2} \le 1\right\}.$$

As a consequence of the Hörmander condition, we can represent any vector in the form $X = \sum_{j=1}^{n} e_j X_j$. The norm $\sqrt{\sum_{j=1}^{m} |e_j|^2}$ is the horizontal norm defined in Definition 1.1. We can extend it as a homogeneous norm on the whole space by setting

$$\|e\| = \left(\sum_{j=1}^{n} |e_j|^{Q/\deg(X_j)}\right)^{1/Q}, \tag{1.9}$$

where Q has been defined in (1.6).

Since the exponential mapping is a local diffeomorphism, we may define:

Definition 1.13. If $\xi_0 \in \Omega$ is fixed, the *canonical coordinates* of ξ around ξ_0 are the coefficients e_j such that

$$\xi = \exp\left(\sum_{j=1}^{n} e_j X_j\right)(\xi_0).$$

This representation will be used to give another characterization of the distance:

Proposition 1.14. *The distance defined in* (1.8) *is locally equivalent to*

$$d_1(\xi, \xi_0) = ||e||,$$

where e are the canonical coordinates of ξ around ξ_0 and $||\,.\,||$ is the homogeneous norm, defined in (1.9).

The proof of this proposition can be found for example in [93], together with a detailed description of properties of the control distance.

1.4.6 Riemannian approximation of the metric

In Definition 1.1, we introduced a horizontal norm only on the horizontal tangent plane. We can extend it to a Riemannian norm on the whole tangent space as follows. For every $\varepsilon > 0$, we define

$$X_j^\varepsilon = \begin{cases} X_j & \text{if } j = 1, \ldots, m, \\ \varepsilon X_j & \text{if } j > m. \end{cases} \tag{1.10}$$

The family X_j^ε, $j = 1, \ldots, n$ formally tends to the family X_j, $j = 1, \ldots, m$ as $\varepsilon \to 0$. We call *Riemannian approximation* of the metric g the Riemannian metric g_ε which makes the vector fields orthonormal. Clearly g_ε restricted to the horizontal plane coincides with the horizontal metric. The geodesic distance associated to g_ε is denoted by d_ε, while the ball in this metric of center ξ_0 and radius r will be denoted by

$$B_\varepsilon(\xi_0, r) = \{\xi : d_\varepsilon(\xi, \xi_0) < \varepsilon\}. \tag{1.11}$$

The distance d_ε tends to the distance d defined in (1.8) as ε goes to 0. We refer to [27] and the references therein for a complete treatment of this topic.

1.4.7 Geodesics and elastica

The curves which minimize the distance are called *geodesics*. We refer to the book of Montgomery [89] for this topic. We do study this problem here but we only recognize the relation between geodesics of $EO(2)$ and elastica. A 2D curve $\widetilde{\gamma} = x(t)$ can be represented in arc length coordinates

$$x'(t) = (\cos(\theta(t)), \sin(\theta(t)))$$

at every point, where θ denotes the direction of the curve at the point $x(t)$. In Subsection 1.4.2 we lifted it to a 3D curve $\gamma(t) = (x(t), \theta(t))$. By the properties of the arc length parametrization,

$$\theta' = k,$$

where k is the Euclidean curvature of $\widetilde{\gamma}$.

The length of the lifted curve is

$$\int \sqrt{(x')^2 + (\theta')^2} = \int \sqrt{(x')^2}\,\sqrt{1 + k^2}.$$

We see that the length of γ is the elastica functional evaluated on $\widetilde{\gamma}$. In this sense, this model can be considered as a neurological motivation of the existing higher-order models of modified elastica (see Subsection 1.2.2).

1.5   Activity propagation and differential operators in Lie groups

1.5.1   Integral curves, association fields, and the experiment of Bosking

Let us go back to the problem of the description of the cortex. Up to now, we have built up a geometric space inspired by the functional geometry of the primary visual cortex. Let us focus on the model in the group $EO(2)$. In the sub-Riemannian space of the cortex, neural activity develops and propagates itself. For the sake of simplicity, in this study we consider an extremely simple model of activity propagation, i.e., a simple linear diffusion along the integral curves of the structure.

This integrative process allows to connect local tangent vectors to form integral curves and is at the base of both regular contour and illusory contour formation [102].

Contour formation has been described by the association field (Field–Hayes–Hess [54]). The anatomical network of horizontal long-range connections has been proposed as the implementation of association fields, and the experiments of Bosking (see Subsection 1.3.3) prove that the diffusion of a marker in the cortex are in perfect agreement with the association fields.

We propose to interpret these lines as a family of integral curves of the generators of $EO(2)$ – the vector fields X_1 and X_2 – starting at a fixed point $\xi = (x, \theta)$:

$$\gamma'(t) = X_1 I(\gamma) + k X_2 I(\gamma), \quad \gamma(0) = (x, \theta),$$

obtained by varying the parameter k in $\mathbb{R}$ (Fig. 1.15).

Long-range connections can consequently be modeled as admissible curves with piecewise constant coefficients k.

1.5.2   Differential calculus in a sub-Riemannian setting

In order to describe the diffusion of the visual signal, we need to recall the main instruments of differential calculus in a sub-Riemannian setting. These properties are well known and can be found for example in the book [17].

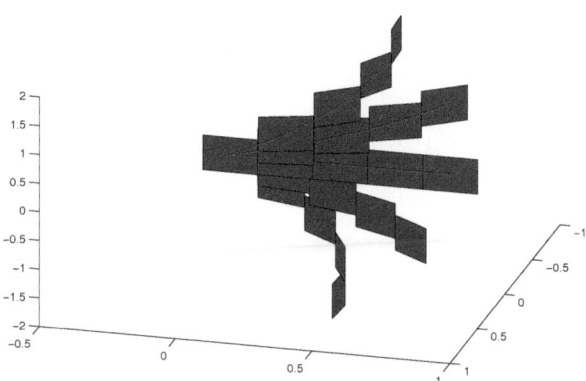

Figure 1.15: Integral curves of the sub-Riemannian structure (compare with the Field–Hayes–Hess association field, Fig. 1.2).

Definition 1.15. Let X be a fixed vector field. We call *Lie derivative* of f in the direction of the vector X on the tangent space to $\mathbb{R}^n$ at a point ξ the derivative with respect to t of the function $f \circ \exp(tX)(\xi)$ at $t = 0$.

Clearly, if f is C^1, then the Lie derivative Xf coincides with the directional derivative, but the Lie derivative can exist even when the directional derivative does not.

Definition 1.16. Let $\Omega \subset \mathbb{R}^n$ be an open set. Let (X_j), $j = 1, \ldots, m$ be a family of smooth vector fields defined on Ω, and let $f \colon \Omega \to \mathbb{R}$. If the Lie derivatives $X_j f$ exist for $j = 1, \ldots, m$, we call *horizontal gradient* of a function f the vector field

$$\nabla_H f = (X_1 f, \ldots, X_m f).$$

A function f is of class C^1_H if $\nabla_H f$ is continuous with respect to the distance defined in (1.8). A function f is of class C^2_H if $\nabla_H f$ is of class C^1_H, and by induction all C^k_H classes are defined.

Note that a C^1_H function need not be differentiable with respect to X_j if $j > m$. It follows that a function of class C^1_H need not be of class C^1_E in the standard Euclidean sense. If the vector fields (X_j), $j = 1, \ldots, m$ have step s, then a function f of class C^s_H is C^1_E.

Remark 1.17. If the vector fields (X_j), $j = 1, \ldots, m$ satisfy the Hörmander condition, then f is C^∞_H if and only if it is of class C^∞_E in the standard sense.

Remark 1.18. The Heisenberg group and the group $EO(2)$ with the choice of vector fields made in Examples 1.6 and 1.7 are of step 2. Hence, if a function f

is of class C_H^k in one of these structures, then it is of class $C_E^{k/2}$ in the standard sense.

From the definition of Lie derivative and the properties of integral curves, the next result follows:

Proposition 1.19. *Let $\Omega \subset \mathbb{R}^n$. Let X and Y be horizontal vector fields defined on Ω and let $f \colon \Omega \to \mathbb{R}$. Assume that at every point ξ in Ω the Lie derivatives $Xf(\xi)$ and $Yf(\xi)$ exist and are continuous. If $\gamma(t) = \exp(tX)(\exp(tY)(\xi))$, then*

$$(f \circ \gamma)'(0) = Xf(\xi) + Yf(\xi) \quad exists.$$

Proof. By the mean value theorem,

$$\frac{1}{t}\Big(f(\gamma(t)) - f(\gamma(0))\Big)$$
$$= \frac{1}{t}\Big(f(\exp(tX)(\exp(tY)(\xi))) - f(exp(tY)(\xi))\Big) + \frac{1}{t}\Big(f(\exp(tY)(\xi)) - f(\xi)\Big)$$
$$= Xf(\exp(t_1 X)(\exp(tY)(\xi))) + Yf(\exp(t_2 Y)(\xi)) \longrightarrow Xf(\xi) + Yf(\xi)$$

as $t \to 0$. $\qquad\qquad\qquad\qquad\qquad\qquad\qquad\qquad\qquad\qquad\qquad\qquad\qquad\qquad\square$

The next fact follows immediately from the previous proposition.

Remark 1.20. If C is the function defined in Lemma 1.12,

$$C(t) = \exp(-tY)\exp(-tX)\exp(tY)\exp(tX)(\xi),$$

and if $f \in C_H^1(\Omega)$ then

$$\frac{d}{dt}(f \circ C)(0) = 0 \quad exists.$$

Proposition 1.21. *Let $\Omega \subset \mathbb{R}^n$, and assume that a collection of vector fields (X_j), $j = 1, \ldots, m$ is defined on Ω and satisfies the Hörmander condition (see Definition 1.4). If f is of class $C_H^1(\Omega)$, then*

(a) *f is continuous in Ω.*

(b) *If C_p is the function defined in (1.7), then the function f satisfies*

$$f(C_p(e)(\xi)) - f(\xi) = \sum_{j=1}^{m} e_j X_j + o(||e||)$$

as $||e|| \to 0$, where $||\,.\,||$ is the homogeneous norm defined in (1.9).

The second assertion is a direct consequence of the previous remark and proposition, together with the definition of C_p. The fact that f is continuous follows from the fact that C_p is a local diffeomorphism; see the proof of connectivity.

Proposition 1.22. *Let $\Omega \subset \mathbb{R}^n$ and let $f \colon \Omega \to \mathbb{R}$ be a continuous function for which the Lie derivatives Xf and Yf exist and are continuous. Then*

$$(X + Y)f = Xf + Yf \quad \text{also exists in } \Omega.$$

Proof. Arguing as in Lemma 1.12, we immediately see that

$$|\exp(tX)\exp(tY)(\xi) - \exp(t(X + Y))(\xi)| = O(t^2),$$

locally uniformly in ξ. It follows that

$$\frac{1}{t}\Big(f(\exp(t(X + Y))(\xi)) - f(\xi)\Big)$$
$$= \frac{1}{t}\Big(f(\exp(tX)(\exp(tY)(\xi))) - f(\xi)\Big) + O(t) \longrightarrow Xf(\xi) + Yf(\xi)$$

as $t \to 0$, by Proposition 1.19. $\qquad\square$

Definition 1.23. Let Ω be an open set in $\mathbb{R}^n$ and assume that a family of vector fields X_j, $j = 1, \ldots, m$ is defined on Ω and satisfies the Hörmander condition. A function $f \colon \Omega \to \mathbb{R}$ is *differentiable at a point* $\xi \in \Omega$ *in the intrinsic sense* if

$$f\Big(\sum_{j=1}^{n} \exp(e_j X_j)(\xi)\Big) - f(\xi) = \sum_{j=1}^{m} e_j X_j f(\xi) + o(\|e\|)$$

as $\|e\| \to 0$. Note that only vector fields of degree 1 appear in the definition.

As a direct consequence of the previous propositions, we have:

Proposition 1.24. *Let $\Omega \subset \mathbb{R}^n$, and assume that a collection of vector fields X_j, $j = 1, \ldots, m$ is defined on Ω and satisfies the Hörmander condition. If f is of class $C_H^1(\Omega)$, then it is differentiable.*

The previous result implies in particular the following:

Remark 1.25. Let $\Omega \subset \mathbb{R}^n$ be an open set and let f be a function in $C_H^1(\Omega)$. If $\gamma(t) = \exp\Big(\sum_{j=1}^{n} t^{\deg(X_j)} e_j X_j\Big)(\xi_0)$, then

$$\lim_{t \to 0} \frac{f(\gamma(t)) - f(\gamma(0))}{t} = \sum_{j=1}^{m} e_j X_j f(\xi_0)$$

locally uniformly on Ω and with respect to e.

1.5.3 Sub-Riemannian differential operators

Definition 1.26. If $\phi = (\phi_1, \ldots, \phi_m)$ is a C_H^1 section of the horizontal tangent plane, we call *divergence* of ϕ the function

$$\operatorname{div}_H(\phi) = \sum_{j=1}^{m} X_j^* \phi_j,$$

where X_j^* is the formal adjoint of the vector field X_j.

From now on we will assume that the vector field X_j is self adjoint for every j. Accordingly, we will define the *sublaplacian operator* as

$$\Delta_H = \mathrm{div}_H(\nabla_H).$$

A uniformly subelliptic operator mimics the structure of uniformly elliptic operators. An $m \times m$ matrix (A_{ij}) is a *uniformly elliptic matrix* if there exist two real numbers λ and Λ such that

$$\lambda|\xi|^2 \leq \sum_{j=1}^{m} A_{ij}\,\xi_i\xi_j \leq \Lambda|\xi|^2.$$

Accordingly, the operator

$$L_A = \sum_{ij=1}^{m} A_{ij}\,X_iX_j \tag{1.12}$$

is called *uniformly subelliptic*.

We define the *subcaloric equation* (the natural analogue of the heat equation) expressed in terms of the subelliptic operator as

$$\partial_t = L_A.$$

Example 1.27. Note that the solution of a sum of squares of two vector fields in $\mathbb{R}^3$ is not regular in general. Indeed, any function of the variable ξ_3 is a solution of

$$\partial_1^2 + \partial_2^2 = 0 \text{ in } \mathbb{R}^3.$$

We are now ready to state the well-known theorem on hypoellipticity due to Hörmander (see [71]):

Theorem 1.28 (Hörmander Theorem). *If $X_1, \ldots, X_m$ satisfy the Hörmander rank condition, then the associated subelliptic operator and the heat operator are hypoelliptic operators.*

These operators admit a fundamental solution Γ of class C^∞. Existence and local estimates of the fundamental solution in terms of the control distance have been first proved by Folland–Stein [57], Rothschild–Stein [107], and Nagel–Stein–Weinger [93].

They precisely proved that the fundamental solution can be locally estimated as

$$|\Gamma(\xi, \xi_0)| \leq C \frac{d^2(\xi, \xi_0)}{|B(x, d(\xi, \xi_0))|},$$

for all ξ, ξ_0 in a neighborhood of a fixed point and for a suitable constant C. Local and global Gaussian estimates of the fundamental solution have been investigated by many authors. We refer to the book [17] for an exhaustive presentation of the topic.

In the application to the cortex, it is necessary to study elliptic regularization of this type of operators. This means that the vector fields X_j will be replaced by the vectors X_j^ε introduced in (1.10). The matrix A_{ij} will be extended to an $n \times n$ matrix A_{ij}^ε uniformly elliptic in such a way that $A_{ij}^\varepsilon \to A_{ij}$ as $\varepsilon \to 0$. Then the Riemannian approximating operator of the operator (1.12) is

$$L_\varepsilon = \sum_{i,j=1}^n A_{ij}^\varepsilon X_i^\varepsilon X_j^\varepsilon.$$

This operator is clearly uniformly elliptic in Ω, but the ellipticity constant tends to $+\infty$ with ε, since the limit operator is not elliptic. However, for the fundamental solution of this operator it is possible to prove subelliptic estimates uniform in ε (see [36]).

Theorem 1.29. *For every compact set $K \subset \Omega$ and every choice of vector fields in the basis $X_{j_1}^\varepsilon, \ldots, X_{j_k}^\varepsilon$, there exist two positive constants C, C_k independent of ε such that, for all $\xi, \xi_0 \in K$ with $\xi \neq \xi_0$,*

$$|X_{j_1}^\varepsilon \cdots X_{j_k}^\varepsilon \Gamma_\varepsilon(\xi, \xi_0)| \leq C_k \frac{d_\varepsilon^{2-k}(\xi, \xi_0)}{|B_\varepsilon(\xi, d_\varepsilon(\xi, \xi_0))|},$$

where $B_\varepsilon(\xi, r)$ is a ball in the approximating Riemannian metric defined in (1.11).

This theorem provides uniform estimates of the fundamental solution of an operator in terms of its control distance. Letting ε go to 0, it allows to deduce, from regularity results known in the elliptic case, analogous results for the subelliptic situation. In general, this approach allows to work with smooth solutions of an elliptic problem $L_\varepsilon u_\varepsilon = f$ in order to obtain uniform estimates for the limit equation.

A first consequence of this result is the regularity in the intrinsic Sobolev spaces. Let $\Omega_0 \subset \Omega$, and $W_\varepsilon^{k,p}(\Omega_0)$ be the set of functions $f \in L^p(\Omega_0)$ such that

$$X_{i_1}^\varepsilon \cdots X_{i_k}^\varepsilon f \in L^p(\Omega_0), \quad i_1, \ldots, i_k \in \{1, \ldots, n\},$$

with natural norm

$$\|f\|_{W_\varepsilon^{k,p}(\Omega_0)} = \sum_{i_1, \ldots, i_k \in \{1, \ldots, n\}} \|X_{i_1}^\varepsilon \cdots X_{i_k}^\varepsilon f\|_{L^p(\Omega_0)}.$$

Let us give some example of applications. Assume that Q is the homogeneous dimension of the limit operator. Then the following Sobolev-type inequality holds:

Corollary 1.30. *If $u \in W_\varepsilon^{1p}$ and is compactly supported in an open set Ω, then there exists a constant C independent of ε such that*

$$\|u\|_{L^r(\Omega)} \leq C\|u\|_{W_\varepsilon^{k,p}(\Omega)},$$

where $r = Qp/(Q - kp)$.

Corollary 1.31. *Assume that* $u \in L^q_{loc}(\Omega)$ *is a solution of the equation* $L_\varepsilon u = f$ *in* Ω *with* $f \in W^{p,q}_{\varepsilon,X}(\Omega)$, *and let* $K_1 \subset\subset K_2 \subset\subset \Omega$. *Then there exists a constant* C *independent of* ε *such that*

$$||u||_{W^{p+2,q}_{\varepsilon,X}(K_1)} \leq C||f||_{W^{p,q}_{\varepsilon,X}(K_2)}$$

for every $p \geq 1$.

1.6 Regular surfaces in a sub-Riemannian setting

1.6.1 Maximum selectivity and lifting images to regular surfaces

Let us go back to the model of the visual cortex. The mechanism of non-maxima suppression does not lift each level line independently, but is applied to the whole image. If O is the output of the simple cells, the maximum of O over the fiber is attained:

$$|O(x,\bar\theta)| = \max_\theta |O(x,\theta)|.$$

In this process, each point x in the 2D domain of the image is lifted to the point $(x, \bar\theta(x))$, and the whole image domain is lifted to the graph of the function $\bar\theta$:

$$\Sigma = \{(x,\theta) : \theta = \bar\theta(x)\}.$$

This lifted set corresponds to the maximum of activity of the output of the simple cells. Setting $f(x,\theta) = \partial_\theta O(x,\theta)$ and considering only strict maxima, the surface becomes

$$\Sigma = \{(x,\theta) : f(x,\theta) = 0, \ \partial_\theta f(x,\theta) > 0\},$$

where the vector ∂_θ is a horizontal vector.

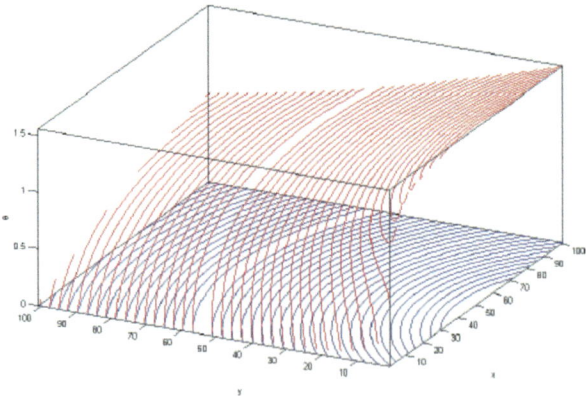

Figure 1.16: Lifting of level lines of an image.

We recall that on the domain of $\bar{\theta}$ only one vector field was defined (see (1.3)):

$$X_{\bar{\theta}} = \cos(\bar{\theta}(x))\partial_1 + \sin(\bar{\theta}(x))\partial_2,$$

which is tangent to the level lines of I.

We will see that Σ is a regular surface in the sub-Riemannian structure, and that in any sub-Riemannian structure the implicit function $\bar{\theta}$ is regular with respect to non-linear vector fields depending on $\bar{\theta}$.

1.6.2 Definition of a regular surface

In this setting, the notion of regular surface is not completely clear. The first definition, given by Federer in [53], was that a regular surface is the image of an open set of $\mathbb{R}^{n-1}$ through a Lipschitz continuous function. However, the Heisenberg group turns out to be completely non rectifiable in this sense [1]. A more natural definition of regular surface has been given by Franchi, Serapioni and Serra Cassano, and investigated in a long series of papers [58–61].

Definition 1.32. A *regular surface* is a subset Σ of $\mathbb{R}^n$ that can be locally represented as the zero-level set of a function $f \in C_H^1$ such that $\nabla_H f(\xi) \neq 0$. In this case, the vector

$$\nu_H = \frac{\nabla_H f(\xi)}{|\nabla_H f(\xi)|}$$

is called *intrinsic normal* of Σ.

The vector ν_H takes the place of a normal vector to the surface. It can be recovered through a blow up procedure similar to the De Giorgi method for the Euclidean proof of rectifiability. We refer to [58] for the proof in the Heisenberg setting and to [34] for the proof in a general setting.

If the vector $\nabla_H f(\xi)$ vanishes at a point ξ, this point is called *characteristic*.

Example 1.33. The generators of the Heisenberg algebra introduced in (1.2) are

$$X_1 = \partial_1 + u\partial_2, \qquad X_2 = \partial_u$$

in $\mathbb{R}^3$, whose points are denoted by $\xi = (x, u)$. The plane $u = 0$ has as intrinsic normal $\nu_H = \partial_u$, which never vanishes, so that the plane is a regular surface. The intrinsic normal of the plane $y = 0$ is uX_1. Hence the points $(x_1, x_2, 0)$ are characteristic for this plane.

Example 1.34. We provide an example of a characteristic surface in the group $EO(2)$, defined in Example 1.7. The points of the space will be denoted by (x, θ) as before. Let us denote by $\widetilde{\gamma}$ a curve in the plane x and let us consider the surface

$$\Sigma = \{(x, \theta) : x \in \widetilde{\gamma}, \theta \in [0, 2\pi]\}.$$

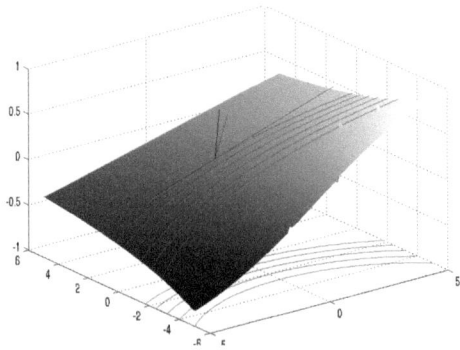

Figure 1.17: A regular surface, foliated
in horizontal curves.

In Subsection 1.4.7 we pointed out that the lifting of the curve $\widetilde{\gamma}$ is a new curve γ whose tangent vector is $X_1 + kX_2$, where k is the Euclidean curvature of $\widetilde{\gamma}$. Hence at every point of the lifted curve γ the surface Σ has two horizontal tangent vectors: $X_1 + kX_2$ and X_2. Consequently, all these points are characteristic.

1.6.3   Implicit function theorem

Regular surfaces in this setting are not regular in the Euclidean sense. An example of an intrinsic regular surface which has a fractal structure was provided by Kirchheim and Serra Cassano [76]. However, a first proof of the Dini theorem for hypersurfaces was given by Franchi, Serapioni and Serra Cassano [58] in the Heisenberg group. A much simpler proof in a general sub-Riemannian structure was given by Citti and Manfredini [35]. Indeed, due to the structure of the vector fields, the implicit function u found in [58] is not a graph in the standard sense. The problem is related to the fact that the definition of a graph is not completely intrinsic, but it assigns a different role to the first variable, lying in the image of u with respect to the other $n - 1$ variables belonging to the domain of u.

Hence we choose a suitable change of variables. As usual, we write $x \in \mathbb{R}^{n-1}$ for the variables in the domain of the implicit function and $y \in \mathbb{R}$ for the other variable. In this way, we represent the points of the space in the form

$$\xi = (y, x).$$

It is also possible to choose the new variables in such a way that the generators of the Lie algebra are written as follows:

$$X_j = \sum_{k=1}^{n-1} a_{jk}(\xi) \partial_{x_k}, \quad j = 1, \dots, m-1, \quad X_m = \partial_y.$$

Let us note that the explicit expression of the vector fields appearing in the model of the cortex is of this type.

In these new variables from the classical implicit function theorem, we immediately deduce the following:

Lemma 1.35. *Let* $\Omega \subset \mathbb{R}^n$ *be an open set. Let* $0 \in \Omega$ *and* $f \in C_X^1(\Omega)$ *be such that*

$$\partial_y f(0) > 0, \quad f(0) = 0.$$

If $\Sigma = \{\xi \in \Omega : f(\xi) = 0\}$, *then there exist neighborhoods* $I \subset \mathbb{R}^{n-1}$, $J \subset \mathbb{R}$ *of* 0 *and a continuous function* $u \colon I \to J$ *such that*

$$\Sigma \cap (J \times I) = \{(u(x), x) : x \in I\}.$$

Proof. The existence of the function u is standard. We recall here only the proof of the continuity of u, in order to point out that in this part of the proof we only need the continuity of the derivative $\partial_y f$, which is here continuous by assumption, since it is horizontal. By the mean value theorem,

$$
\begin{aligned}
0 &= f(u(x), x) - f(u(x_0), x_0) \\
&= f(u(x), x) - f(u(x_0), x) + f(u(x_0), x) - f(u(x_0), x_0) \\
&= \partial_y f(s, x)(u(x) - u(x_0)) + f(u(x_0), x) - f(u(x_0), x_0).
\end{aligned}
$$

Then

$$|u(x) - u(x_0)| = \left| \frac{f(u(x_0), x) - f(u(x_0), x_0)}{\partial_y f(s, x)} \right| = o(1),$$

since the denominator is bounded away from 0 by assumption, and the function f is continuous. $\square$

In order to study the regularity of the function u, we will need to project the vector fields X_j on its domain. To this end, we define a projection on $\mathbb{R}^{n-1}$,

$$\pi(\xi) = x,$$

and a projection on its tangent plane:

$$\pi_u \left(\sum_{k=1}^{n-1} a_k(\xi) \partial_{x_k} \right) = \sum_{k=1}^{n-1} a_k(u(x), x) \partial_{x_k}.$$

Accordingly, we define

$$X_{ju} = \pi_u(X_j), \quad j = 1, \dots, m-1.$$

In particular, the projection of the elements of the basis will be

$$\pi_u(X_m) = 0$$

and

$$X_{ju} = \sum_{k=1}^{n-1} a_{jk}(u(x), x) \partial_{x_k}, \quad j = 1, \dots, m-1.$$

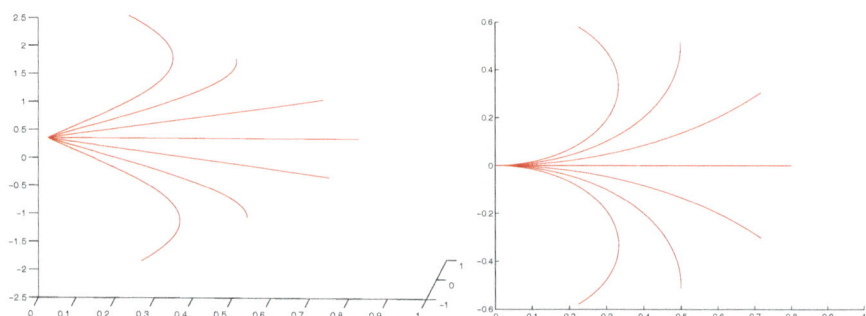

Figure 1.18: Integral curves of the vector fields and their $(n-1)$D projection.

Definition 1.36. Let $I \subset P$ be an open set. We say that a continuous function $u\colon I \to \mathbb{R}$ is of class $C_u^1(I)$ if $X_{ju}(x)$ exists for $j = 1, \ldots, m-1$ and every $x \in I$, and they are continuous. The *intrinsic gradient* is defined as

$$\nabla_u u = (X_{1u}u, \ldots, X_{(m-1)u}u).$$

Theorem 1.37. *If the assumptions of Lemma 1.35 are satisfied, then the implicit function u is of class C_u^1, and*

$$\nabla_u u(x_0) = -\frac{(X_1 f(\xi_0), \ldots, X_{m-1} f(\xi_0))}{\partial_y f(\xi_0)}.$$

Proof. Let us consider the vector X_{ju}, a point x_0, and call $\gamma_u(t) = \exp(tX_{ju})(x_0)$. We also call $\gamma(t) = \exp(tX_j)(u(x_0), x_0)$ and $\gamma_\pi(t) = \pi(\exp(tX_j)(u(x_0), x_0))$. Then, by definition of Σ,

$$\begin{aligned}
0 &= f(u(\gamma_u(t)), \gamma_u(t)) - f(u(\gamma_u(0)), \gamma_u(0)) \\
&= f(u(\gamma_u(t)), \gamma_u(t)) - f(u(\gamma_u(0)), \gamma_u(t)) + f(u(\gamma_\pi(0)), \gamma_u(t)) \\
&= -f(u(\gamma_\pi(0)), \gamma_\pi(t)) + f(u(\gamma_\pi(0)), \gamma_\pi(t)) - f(\gamma(0)).
\end{aligned}$$

By the classical mean value theorem, there exist z and c such that the last expression is equal to

$$\partial_y f(c, \gamma_u(t)) \left(u(\gamma_u(t)) - u(\gamma_\pi(0)) \right) + (X_1, \ldots, X_{m-1}) \, \partial_y f(u(\gamma_\pi(0)), z)$$
$$(\gamma_u(t) - \gamma_\pi(t)) - (f \circ \gamma)(t) - (f \circ \gamma)(0).$$

Note that the first component of the curve γ is constant, and therefore $\gamma(t) = (u(\gamma_\pi(0)), \gamma_\pi(t))$. Dividing by t and letting t go to 0, we obtain

$$0 = \partial_y f(\xi_0) X_{ju} u(x_0) + X_j f(\xi_0).$$

Therefore,

$$X_{ju}u(x_0) = -\frac{X_j f(\xi_0)}{\partial_y f(\xi_0)}. \qquad \square$$

1.6.4 Non-regular and non-linear vector fields

A consequence of the Dini theorem is the fact that, if we start with a regular surface of class C_H^1, its implicit function u is differentiable with respect to the non-linear vector fields (X_{ju}). This opens a large spectrum of problems, since these new vector fields are non regular, and in general they satisfy conditions different from the initial vector fields.

Let us give some examples.

Example 1.38. Let us consider a Heisenberg group of higher dimension. This is $\mathbb{R}^5$ with the choice of vector fields

$$X_1 = \partial_1 + \frac{\xi_2}{2}\partial_5, \quad X_2 = \partial_2 - \frac{\xi_1}{2}\partial_5, \quad X_3 = \partial_3 + \frac{\xi_4}{2}\partial_5, \quad X_4 = \partial_4 - \frac{\xi_3}{2}\partial_5.$$

Since

$$[X_2, X_1] = \partial_5, \qquad (1.13)$$

these vector fields satisfy the Hörmander rank condition. With the change of variables introduced in the previous subsection, these operators become

$$X_1 = \partial_{x_1} + x_2\partial_{x_4}, \quad X_2 = \partial_{x_2}, \quad X_3 = \partial_{x_3} + y\partial_{x_4}, \quad X_4 = \partial_y$$

in $\mathbb{R}^5$. The associated non-linear vector fields in the tangent space to $\mathbb{R}^4$ are

$$X_{1u} = \partial_{x_1} + x_2\partial_{x_4}, \quad X_2 = \partial_{x_2}, \quad X_3 = \partial_{x_3} + u\partial_{x_4}$$

in $\mathbb{R}^4$. It is clear that, if u is smooth, then these are Hörmander vector fields, by condition (1.13). However, in general, the solution u will be only C_u^1, and the difficulty in handling these vectors is the lack or regularity. We say that a *weak Hörmander condition* is verified.

In this situation, there is reasonable hope to prove the Poincaré inequalities – estimates of the fundamental solution – and mimic in this non-regular situation results that are known in the smooth setting. A first Poincaré inequality for non-regular vector fields was established in [79]. After that, an inequality of this type was proved in [86] for vector fields of class C^2 and step 2. A similar inequality requires C^{s+1} regularity for vector fields of step s [21, 87]. Very recently, a Poincaré inequality for Heisenberg non-linear vector fields of class C^1 has been proved by Manfredini in [80]. From this, a Sobolev inequality with optimal exponent follows. Estimates for the fundamental solution for non-linear vector fields have been proved in [81].

Example 1.39. In the case of the Heisenberg group of dimension 1 (see Example 1.6), we have a Lie algebra with two generators in a 3D space. The vector fields X_1, X_2 projected on the plane x reduce to only one vector field:

$$X_{1u} = \partial_{x_1} + u(x)\partial_{x_2}. \tag{1.14}$$

In this case, we have a unique non-linear vector field in $\mathbb{R}^2$. It is clear that this vector field does not satisfy the Hörmander condition, not even when u is smooth.

In this low-dimensional case, a few results are known only for the Heseinberg group. More recently, Ambrosio, Serra Cassano and Vittone gave a characterization of implicit functions in [3], while Bigolin and Serra Cassano started the study of the set of C_u^1 functions in [16]. In this case, there is no hope to prove an estimate of the fundamental solution of linear operators defined in terms of non-linear vector fields. For these operators, the Riemannian approximation can be extremely useful. Indeed, using the estimate of the approximating fundamental solution, Citti, Capogna and Manfredini proved a Sobolev estimate for the linearized operator,

$$\sum_{ij} A_{ij} X_{ju}^\varepsilon X_{ju}^\varepsilon z = 0, \tag{1.15}$$

where A_{ij} is positive definite and

$$X_{1u}^\varepsilon = X_{1u}, \quad X_{2u}^\varepsilon = \varepsilon\partial_{x_2}, \quad \nabla_u^\varepsilon = (X_{1u}^\varepsilon, X_{2u}^\varepsilon). \tag{1.16}$$

The result in [24] reads as follows:

Theorem 1.40. *Let us assume that z is a classical solution of the approximated problem* (1.15), *where u is a smooth function. Assume that there exists a constant C independent of ε such that*

$$||A_{ij}||_{C^\alpha(K)} + ||u||_{C^{1,\alpha}(K)} + ||\partial_2 z||_{L^p(K)} + ||\partial_2 X_u z||_{L^q(K)} + ||(\nabla_u^\varepsilon)^2 z||_{L^2(K)} \le C.$$

Then for any compact set $K_1 \subset\subset K$ there exists a constant C_1 only dependent on K and C such that

$$||z||_{W_\varepsilon^{2,r}(K_1)} \le C_1,$$

where $r = \min(5q/(5 - (1 + \alpha)q), 5p/(5 - \alpha p))$.

The proof is based on the estimates of the fundamental solution uniform in ε stated in Theorem 1.29. The exponent r is reminiscent of a Sobolev exponent, modeled on a homogeneous dimension $Q = 5$. However, it is not optimal, since the coefficients are not regular.

1.7   Completion and minimal surfaces

1.7.1   A completion process

The joint work of sub-Riemannian diffusion (Section 1.5) and non-maxima suppression (Section 1.6) allows us to propagate existing information and then to complete boundaries and surfaces. Starting from the lifted surface, the two mechanisms are simultaneously applied until the completion is reached. To take into account the simultaneous work of diffusion and non-maxima suppression, we consider iteratively diffusion in a finite time interval followed by non-maxima suppression, and we compute the limit when the time interval tends to 0.

The algorithm is an extension of the diffusion-driven motion by curvature introduced by Bence, Merriman and Osher in [14]. It is described by induction as follows. Given a function u_n, whose maxima in a given direction are attained on a surface Σ_n, we diffuse in an interval of length h,

$$v_t = \Delta_H v, \quad v_{t=0} = v_{\Sigma_n}, \quad t \in [nh, (n+1)h].$$

At time $(n+1)h$, the solution defines a new function v_{n+1}, and we build a new surface through non-maxima suppression:

$$\Sigma_{n+1}((n+1)h) = \{\partial_{\nu_{\Sigma_n}} v_{n+1} = 0, \; \partial^2_{\nu_{\Sigma_n}} v_{n+1} < 0\}.$$

If we fix a time T, we can choose intervals of length $h = T/(n+1)$, and we get the two sequences $v_{n+1}(\cdot, T)$, $\Sigma_{n+1}(T)$. We expect convergence of the two sequences $\Sigma_n(T)$ and $u_n(T)$ respectively to the mean curvature flow $\Sigma(T)$ of the surface Σ_0 and the Beltrami flow on Σ. For $T \to +\infty$, the function $\Sigma(T)$ should converge to a minimal surface in the rototranslation space, in the sense that its curvature identically vanishes.

The formal proof of the convergence of diffusion-driven motion by curvature in the Euclidean setting is due to Evans [52] and Barles–Georgelin [8]. The proof of the analogous assertion in this context is still work in progress.

By now we have studied properties of minimal surfaces and verified that they have the properties required by the completion model.

1.7.2   Minimal surfaces in the Heisenberg group

Several equivalent notions of *horizontal mean curvature* H_0 for a regular C^2_H surface $M \subset \mathbb{H}^1$ (outside characteristic points) have been given in the literature. To quote a few: H_0 can be defined in terms of the first variation of the area functional [30, 41, 68, 88, 106, 115] as horizontal divergence of the horizontal unit normal. As such, the expression of the curvature of a surface level set of a function f becomes

$$H_0 f = \sum_{j=1}^{m} X_j \left(\frac{X_j f}{|\nabla_H f|} \right).$$

A different but equivalent notion of curvature in terms of a notion of a metric normal has been given by [4]. In [27] it has been recognized that the curvature can be obtained as a limit of the mean curvatures H_ε in the Riemannian metrics g_ε defined in Subsection 1.4.6. The definition of H_ε can be given in terms of the vector fields X_j^ε defined in (1.10), as follows:

$$H_\varepsilon = \sum_{j=1}^n X_j^\varepsilon \left(\frac{X_j^\varepsilon f}{|\nabla_\varepsilon f|} \right).$$

Here ∇_ε denotes the approximated gradient

$$\nabla_\varepsilon = (X_1^\varepsilon \cdots X_n^\varepsilon).$$

In the particular case of intrinsic graphs, it can be expressed in terms of the vector fields (X_{ju}) defined in Section 1.6. As we already noted, the regularity theory for intrinsic minimal surfaces is completely different depending on whether a weak Hörmander-type condition is satisfied or not. In $\mathbb{H}^n$ with $n > 1$, this condition is satisfied and the problem has been afforded in [25].

Hence here we focus on the low-dimensional case, which naturally arises from the application to the visual cortex. By simplicity we restrict to the monodimensional Heisenberg group. The extension to general Lie algebras with two generators, step 2, and dimension 3 is due to [6]. Through the implicit function theorem we have defined in (1.14) a unique vector field X_{1u} on $\mathbb{R}^2$.

The curvature operator for intrinsic graphs reduces to

$$X_{1u} \left(\frac{X_{1u} u}{\sqrt{1 + |X_{1u} u|^2}} \right) = f, \text{ for } x \in \Omega \subset \mathbb{R}^2. \tag{1.17}$$

Properties of regular minimal surfaces have been studied in [9, 30, 32, 42, 65, 66, 95, 98]. The Riemannian approximating vector fields have been defined in (1.16), while the Riemannian approximating operator reads as follows:

$$L_\varepsilon u = \sum_{i=1}^2 X_{iu}^\varepsilon \left(\frac{X_{iu}^\varepsilon u}{\sqrt{1 + |\nabla_u^\varepsilon u|^2}} \right) = f, \text{ for } x \in \Omega \subset \mathbb{R}^2. \tag{1.18}$$

Using this approximation, we can give the definition of vanishing viscosity solution.

Definition 1.41. If C_E^1 denotes the standard Euclidean C^1 norm, we say that an Euclidean Lipschitz continuous function u is a *vanishing viscosity solution* of (1.17) in an open set Ω if there exists a sequence $\varepsilon_j \to 0$ as $j \to +\infty$, and a sequence (u_j) of smooth solutions of (1.18) in Ω, such that, for every compact set $K \subset \Omega$,

- $\|u_j\|_{C_E^1(K)} \leq C$ for every j;
- $u_j \to u$ as $j \to +\infty$ pointwise a.e. in Ω.

Existence of viscosity solutions has been proved by Cheng, Hwang and Yangin in [32], while the problem of regularity of minimal surfaces has been afforded in [24]. This result reads as follows:

Theorem 1.42. *The Lipschitz continuous vanishing viscosity solutions of* (1.17) *are intrinsically smooth functions.*

This theorem highlights a very general idea: any positive semi-definite operator of second order regularizes in the direction of its positive eigenvalues. However, in general, this does not imply smoothness of solutions, since regularity can be expected only in the directions of the non-vanishing eigenvalues. Indeed, the following foliation result holds for minimal graphs:

Corollary 1.43. *Let* $\{x_3 = u(x),\ x \in \Omega\}$ *be a Lipschitz continuous vanishing viscosity minimal graph. The flow of the vector* $X_{1u}u$ *yields a foliation of the domain* Ω *by polynomial curves* γ *of degree two. For every fixed* $x_0 \in \Omega$, *denote by* γ *the unique leaf passing through that fixed point. The function* u *is differentiable at* x_0 *in the Lie sense along* γ *and the equation* (1.17) *reduces to* $\frac{d^2}{dt^2}(u(\gamma(t))) = 0$.

Remark 1.44. To better understand the notion of intrinsic regularity, we consider the non-smooth minimal graph

$$u(x) = \frac{x_2}{x_1 - \operatorname{sgn}(x_2)}.$$

Although this function is not C^1 in the Euclidean sense, observe that $X_{1u}u = 0$ for every $x \in \Omega$. Hence, this is an example of a minimal surface which is not smooth but which can be differentiated indefinitely in the direction of the Legendrian foliation. Another example of a non-regular minimal surface has been provided in [99].

1.7.3 Uniform regularity for the Riemannian approximating minimal graph

In this subsection, we fix a solution of the Riemannian approximating equation and establish a priori estimates, uniform in ε. A complete proof is contained in [24]. We give here a short presentation of the proof.

To this end, we assume that f is a fixed smooth function defined on an open set Ω of $\mathbb{R}^2$, and that u is a solution of (1.18) in Ω. We also assume that

$$M = ||u||_{L^\infty(\Omega)} + ||\nabla_u^\varepsilon u||_{L^\infty(\Omega)} + ||\partial_2 u||_{L^\infty(\Omega)} < \infty. \tag{1.19}$$

The necessary estimates will be provided in suitable Sobolev spaces defined in terms of the vector fields.

Definition 1.45. We say that $\phi \in W_\varepsilon^{1,p}(\Omega)$, $p > 1$, if $\phi, \nabla_u^\varepsilon \phi \in L^p(\Omega)$. In this case, we set

$$||\phi||_{W_\varepsilon^{1,p}(\Omega)} = ||\phi||_{L^p(\Omega)} + ||\nabla_u^\varepsilon \phi||_{L^p(\Omega)}.$$

We say that $\phi \in W_\varepsilon^{k,p}(\Omega)$ if $\phi \in L^p$ and $\nabla_u^\varepsilon \phi \in W_\varepsilon^{k-1,p}(\Omega)$.

If $\varepsilon = 0$, we give an analogous definition of Sobolev spaces in the sub-Riemannian setting. We denote by $W_0^{k,p}(\Omega)$ the space of $L^p(\Omega)$ functions ϕ such that

$$X_{1u}^\varepsilon \phi, (X_{1u}^\varepsilon)^2 \phi, \ldots, (X_{1u}^\varepsilon)^k \phi \in L^p(\Omega).$$

Using in full strength the non-linearity of the operator L_ε, we prove here some Cacciopoli-type inequalities for the intrinsic gradient of u and for the derivative $\partial_2 u$.

We first prove that, if u is a smooth solution of equation (1.18), then its derivatives $\partial_2 u$ and $X_{ku}^\varepsilon u$ are solutions of a new second order equation, defined in terms of vector fields:

$$M_\varepsilon z = \sum_{ij=1}^2 X_{iu}^\varepsilon \left(\frac{A_{ij}(\nabla_u^\varepsilon u)}{\sqrt{1 + |\nabla_u^\varepsilon u|^2}} X_{ju}^\varepsilon z \right), \quad \text{where } A_{ij}(p) = \delta_{ij} - \frac{p_i p_j}{1 + |p|^2}. \quad (1.20)$$

We first observe that

$$\partial_2 X_{iu}^\varepsilon u = -(X_{iu}^\varepsilon)^* \partial_2 u,$$

where $(X_{iu}^\varepsilon)^*$ is the L^2-adjoint of the differential operator X_{iu}^ε and

$$(X_{1u}^\varepsilon)^* = -X_{1u}^\varepsilon - \partial_2 u, \quad (X_{2u}^\varepsilon)^* = -X_{2u}^\varepsilon. \quad (1.21)$$

Lemma 1.46. *If u is a smooth solution of (1.18), then $v = \partial_2 u$ is a solution of the equation*

$$\sum_{i,j} (X_{iu}^\varepsilon)^* \left(\frac{A_{ij}(\nabla_u^\varepsilon u)}{\sqrt{1 + |\nabla_u^\varepsilon u|^2}} (X_{ju}^\varepsilon)^* v \right) = 0, \quad (1.22)$$

where A_{ij} are defined in (1.20). This equation can be equivalently represented as

$$M_\varepsilon v = f_1(\nabla_u^\varepsilon u) v^3 + f_{2,i}(\nabla_u^\varepsilon u) v X_{iu}^\varepsilon v^2 + X_i \left(f_{3,i}(\nabla_u^\varepsilon u) v^2 \right), \quad (1.23)$$

for suitable smooth functions f_1 and $f_{j,i}$. Analogously, the function $z = X_{ku}^\varepsilon u$ with $k \leq 2$ is a solution of the equation

$$M_\varepsilon z = f_1(\nabla_u^\varepsilon u) v^2 + f_{2,i}(\nabla_u^\varepsilon u) X_{iu}^\varepsilon v^2 + X_i \left(f_{3,i}(\nabla_u^\varepsilon u) v \right). \quad (1.24)$$

Proof. Let us prove the first assertion. Differentiating the equation (1.18) with respect to ∂_2, we obtain

$$\partial_2 \left(X_{iu}^\varepsilon \left(\frac{X_{iu}^\varepsilon u}{\sqrt{1 + |\nabla_u^\varepsilon u|^2}} \right) \right) = 0.$$

Using (1.21),

$$(X_{iu}^\varepsilon)^* \left(\partial_2 \left(\frac{X_{iu}^\varepsilon u}{\sqrt{1 + |\nabla_u^\varepsilon u|^2}} \right) \right) = 0.$$

Note that

$$
\partial_2 \left(\frac{X_{iu}^\varepsilon u}{\sqrt{1 + |\nabla_u^\varepsilon u|^2}} \right) = \frac{\partial_2 X_{iu}^\varepsilon u}{\sqrt{1 + |\nabla_u^\varepsilon u|^2}} - \frac{X_{iu}^\varepsilon u\, X_{ju}^\varepsilon u\, \partial_2 X_{ju}^\varepsilon u}{(1 + |\nabla_u^\varepsilon u|^2)^{3/2}}
$$

$$
= -\frac{(X_{iu}^\varepsilon)^* \partial_2 u}{\sqrt{1 + |\nabla_u^\varepsilon u|^2}} + \frac{X_{iu}^\varepsilon u\, X_{ju}^\varepsilon u\, (X_{ju}^\varepsilon)^* \partial_2 u}{(1 + |\nabla_u^\varepsilon u|^2)^{3/2}} = \frac{A_{ij}(\nabla_u^\varepsilon u)}{\sqrt{1 + |\nabla_u^\varepsilon u|^2}} (X_{ju}^\varepsilon)^* v,
$$

so the first assertion is proved.

Assertion (1.23) follows from (1.21) and (1.22). Indeed,

$$
0 = \sum_{i,j} X_{iu}^\varepsilon \left(\frac{A_{ij}(\nabla_u^\varepsilon u)}{\sqrt{1 + |\nabla_u^\varepsilon u|^2}} X_{ju}^\varepsilon v \right) + \sum_i X_{iu}^\varepsilon \left(\frac{A_{i1}(\nabla_u^\varepsilon u)}{\sqrt{1 + |\nabla_u^\varepsilon u|^2}} v^2 \right)
$$

$$
+ \sum_j \frac{A_{1j}(\nabla_u^\varepsilon u)}{\sqrt{1 + |\nabla_u^\varepsilon u|^2}} v X_{ju}^\varepsilon v + \frac{A_{11}(\nabla_u^\varepsilon u)}{\sqrt{1 + |\nabla_u^\varepsilon u|^2}} v^3.
$$

We omit the proof of (1.24), which is a similar direct verification. □

Since the operator M_ε in (1.20) is in divergence form, it is quite standard to prove the following intrinsic Cacciopoli-type inequalities:

Proposition 1.47 (Intrinsic Cacciopoli-type inequality). *Let u be a smooth solution of (1.18), satisfying (1.19). Let us denote*

$$
z = X_{uk}^\varepsilon u + 2M, \quad v = \partial_2 u + 2M,
$$

where M is the constant in (1.19). Then for every p there exists a constant C, only dependent on p and M, such that, for every $\phi \in C_0^\infty$,

$$
\int |\nabla_u^\varepsilon v|^2 z^{p-2} \phi^2 \le C \int z^p (\phi^2 + |\nabla_u^\varepsilon \phi|^2) + \int |\nabla_u^\varepsilon z|^2 z^{p-2} \phi^2,
$$

$$
\int |\nabla_u^\varepsilon z|^2 z^{p-2} \phi^2 \le C \int z^p (\phi^2 + |\nabla_u^\varepsilon \phi|^2).
$$

Proof. Since A_{ij} is uniformly elliptic, we have

$$
\int |\nabla_u^\varepsilon v|^2 z^{p-2} \phi^2 \le C \int \frac{A_{ij}(\nabla_u^\varepsilon u)}{\sqrt{1 + |\nabla_u^\varepsilon u|^2}} X_{iu}^\varepsilon v X_{ju}^\varepsilon v z^{p-2} \phi^2
$$

(using the expression (1.21) of the formal adjoint)

$$
= -C \int \frac{A_{ij}(\nabla_u^\varepsilon u)}{\sqrt{1 + |\nabla_u^\varepsilon u|^2}} (X_{iu}^\varepsilon)^* v\, X_{ju}^\varepsilon v\, z^{p-2}\, \phi^2
$$

$$
+ C \int \frac{A_{1j}(\nabla_u^\varepsilon u)}{\sqrt{1 + |\nabla_u^\varepsilon u|^2}} X_{ju}^\varepsilon v\, v \partial_2 u\, z^{p-2} \phi^2
$$

(integrating by parts X_{ju}^ε in the first integral)

$$= C \int (X_{ju}^\varepsilon)^* \left(\frac{A_{ij}(\nabla_u^\varepsilon u)}{\sqrt{1 + |\nabla_u^\varepsilon u|^2}} (X_{iu}^\varepsilon)^* v \right) v z^{p-2} \phi^2$$

$$+ (p-2)C \int \frac{A_{ij}(\nabla_u^\varepsilon u)}{\sqrt{1 + |\nabla_u^\varepsilon u|^2}} (X_{iu}^\varepsilon)^* v \; v X_{ju}^\varepsilon z \; z^{p-3} \phi^2$$

$$+ 2C \int \frac{A_{ij}(\nabla_u^\varepsilon u)}{\sqrt{1 + |\nabla_u^\varepsilon u|^2}} (X_{iu}^\varepsilon)^* v \; v z^{p-2} \phi X_{ju}^\varepsilon \phi$$

$$+ C \int \frac{A_{i\,1}(\nabla_u^\varepsilon u)}{\sqrt{1 + |\nabla_u^\varepsilon u|^2}} (X_{iu}^\varepsilon)^* v \; v \partial_2 u \; z^{p-2} \phi^2 + C \int \frac{A_{1\,j}(\nabla_u^\varepsilon u)}{\sqrt{1 + |\nabla_u^\varepsilon u|^2}} X_{ju}^\varepsilon v \; v \partial_2 u z^{p-2} \phi^2.$$

The first integral vanishes by Lemma 1.46. In the other integrals we can use that

$$\left| \frac{A_{ij}(\nabla_u^\varepsilon u)}{\sqrt{1 + |\nabla_u^\varepsilon u|^2}} \right| \le 1, \quad |v| \le M, \quad \text{and} \quad |(X_{iu}^\varepsilon)^* v| \le (M^2 + |\nabla_u^\varepsilon v|),$$

where M is defined in (1.19). Then, eventually changing the constant C,

$$\int |\nabla_u^\varepsilon v|^2 z^{p-2} \phi^2 \le C \left(\int |\nabla_u^\varepsilon v| |\nabla_u^\varepsilon z| z^{p-3} \phi^2 + \int |\nabla_u^\varepsilon v| z^{p-2} (\phi^2 + |\phi \nabla_u^\varepsilon \phi|) \right)$$

(by Hölder's inequality and the fact that z is uniformly bounded away from 0)

$$\le \delta \int |\nabla_u^\varepsilon v|^2 z^{p-2} \phi^2 + C(\delta) \int |\nabla_u^\varepsilon z|^2 z^{p-2} \phi^2 + C(\delta) \int z^p (\phi^2 + |\nabla_u^\varepsilon \phi|^2).$$

For δ sufficiently small, this implies that

$$\int |\nabla_u^\varepsilon v|^2 z^{p-2} \phi^2 \le C \int |\nabla_u^\varepsilon z|^2 z^{p-2} \phi^2 + C \int z^p (\phi^2 + |\nabla_u^\varepsilon \phi|^2).$$

This proves the first inequality. We omit the proof of the second, which is completely analogous and can be found in [24].   □

We want to prove the $C^{1\alpha}$ regularity of z. The classical proof is based on the Moser procedure. This method requires two ingredients: the Sobolev embedding and the Cacciopoli inequality. Here we have proved an intrinsic Cacciopoli-type inequality, but we cannot prove the intrinsic Sobolev embedding for vector fields with non-regular coefficients. This is why we will establish now an Euclidean Cacciopoli inequality, and use the standard Euclidean procedure for a first gain of regularity:

Proposition 1.48. *Let u be a solution of equation* (1.18) *satisfying* (1.19). *For every compact set $K \subset\subset \Omega$ there exist a real number α and a constant C, only dependent on the constant M in* (1.19), *such that*

$$\|u\|_{W_\varepsilon^{2,2}(K)} + \|\partial_2 u\|_{W_\varepsilon^{1,2}(K)} + \|u\|_{C_u^{1,\alpha}(K)} \leq C.$$

Proof. The first part of the thesis,

$$\|u\|_{W_\varepsilon^{2,2}(K)} + \|\partial_2 u\|_{W_\varepsilon^{1,2}(K)} \leq C,$$

is proved in Proposition 1.47. Let us now establish an Euclidean Cacciopoli-type inequality for $z = X_{ku}^\varepsilon u$. We observe that the Euclidean gradient can be estimated as follows:

$$|\nabla_E z|^2 \leq |X_{1u}^\varepsilon z - u\partial_2 z|^2 + |\partial_2 z|^2 \leq |X_{1u}^\varepsilon z|^2 + C|\partial_2(X_{1u}^\varepsilon u)|^2$$

$$= |X_{1u}^\varepsilon z|^2 + C|(X_{1u}^\varepsilon)^* v|^2 \leq |\nabla_u^\varepsilon z|^2 + |\nabla_u^\varepsilon v|^2 + C.$$

From Proposition 1.47 it follows that for every $p \neq 1$ there exists a constant C only dependent on p such that, for every $\phi \in C_0^\infty$,

$$\int |\nabla_E z|^2 z^{p-2} \phi^2 \leq C \int z^p(\phi^2 + |\nabla_E \phi|^2).$$

Now the thesis follows via the classical Euclidean Moser technique. $\square$

With this better regularity of the coefficients, we can use the Sobolev-type Theorem 1.40 for vector fields with $C^{1,\alpha}$ coefficients to obtain a further gain of regularity.

Proposition 1.49. *Let u be a solution of equation* (1.18) *satisfying* (1.19). *For every compact set $K \subset\subset \Omega$ there exist a real number α and a constant C, only dependent on the constant M in* (1.19), *such that*

$$\|u\|_{W_\varepsilon^{2,10/3}(K)} + \|\partial_2 u\|_{W_\varepsilon^{1,2}(K)} + \|u\|_{C_u^{1,\alpha}(K)} \leq C. \tag{1.25}$$

Proof. We first note that equation (1.18) can be as well written in divergence form,

$$L_\varepsilon = \sum_{ij} A_{ij}(\nabla_u^\varepsilon u) X_{iu}^\varepsilon X_{ju}^\varepsilon,$$

where A_{ij} are the coefficients defined in (1.20). Since the function u satisfies uniform $C_u^{1\alpha}$ estimates, the coefficients $A_{ij}(\nabla_u^\varepsilon u)$ satisfy uniform C^α estimates. Then we can apply Theorem 1.40 using the fact that, for every p,

$$\|\partial_2 u\|_{L^p(K)} + \|\nabla_u^\varepsilon \partial_2 u\|_{L^2(K)} \leq C.$$

It follows that

$$\|u\|_{W_\varepsilon^{2,r}(K)} \leq C,$$

where $r = 10/(5 - 2(1 - \alpha))$. Since we do not have an estimate for α, we will set $\alpha = 0$ and obtain $r = 10/3$. $\square$

Due to the fact that our Sobolev inequality is not optimal, we will also need an interpolation property, which is completely intrinsic and can take the place of a Sobolev inequality:

Proposition 1.50. *For every $p \geq 3$ and every function $z \in C^{\infty}(\Omega)$ there exists a constant C_p, dependent on p and the constant M in (1.19), such that, for every $\phi \in C_0^{\infty}(\Omega)$ and every $\delta > 0$,*

$$\int |X_{iu}^{\varepsilon} z|^{p+1} \phi^{2p} \leq C \int \left(z^{p+1} \phi^{2p} + z^2 |X_{iu}^{\varepsilon} z|^{p-1} \phi^{2p-2} |X_{iu}^{\varepsilon} \phi|^2 \right)$$

$$+ C \int |(X_{iu}^{\varepsilon})^2 z|^2 |X_{iu}^{\varepsilon} z|^{p-3} z^2 \phi^{2p},$$

where i can be either 1 or 2.

Proof. We have

$$\int |X_{iu}^{\varepsilon} z|^{p+1} \phi^{2p} = \int X_{iu}^{\varepsilon} z |X_{iu}^{\varepsilon} z|^p \mathrm{sign}(X_{iu}^{\varepsilon} z) \phi^{2p}$$

(integrating by parts, using (1.21) and the Kronecker function δ_{ij})

$$= -\delta_{1i} \int \partial_2 u z |X_{iu}^{\varepsilon} z|^p \mathrm{sign}(X_{iu}^{\varepsilon} z) \phi^{2p} - p \int z (X_{iu}^{\varepsilon})^2 z |X_{iu}^{\varepsilon} z|^{p-1} \phi^{2p}$$

$$- 2p \int z |X_{iu}^{\varepsilon} z|^p \mathrm{sign}(X_{iu}^{\varepsilon} z) \phi^{2p-1} X_{iu}^{\varepsilon} \phi$$

(by Hölder's inequality)

$$\leq \frac{C}{\delta} \int \left(z^{p+1} \phi^{2p} + z^2 |X_{iu}^{\varepsilon} z|^{p-1} \phi^{2p-2} |X_{iu}^{\varepsilon} \phi|^2 \right)$$

$$+ \delta \int |X_{iu}^{\varepsilon} z|^{p+1} \phi^{2p} + \frac{C}{\delta} \int z^2 |(X_{iu}^{\varepsilon})^2 z|^2 |X_{iu}^{\varepsilon} z|^{p-3} \phi^{2p},$$

and choosing δ sufficiently small we obtain the desired inequality. $\square$

The next step is to iterate the previous argument and obtain higher integrability of the Hessian of u. The proof goes as before: We establish two intrinsic Cacciopoli-type inequalities, for the derivatives of $z = X_{iu}^{\varepsilon} \nabla_u^{\varepsilon} u$ and $v = \partial_2 \nabla_u^{\varepsilon} u$; from this we deduce that u belongs to a better class of Hölder continuous functions; then the intrinsic Sobolev inequality Theorem 1.40 yields the desired estimate of the second derivatives in the natural Sobolev spaces.

Lemma 1.51. *Let $p \geq 3$ be fixed, let $f \in C^{\infty}(\Omega)$, let u be a function satisfying the bound (1.19), and let z be a smooth solution of the equation $M_{\varepsilon} z = f$. Then there* exists a constant C, which depends on p and on the constant M in (1.19), but is independent of ε and z, such that, for every $\phi \in C_0^\infty(\Omega)$, $\phi > 0$,

$$
\int |\nabla_u^\varepsilon(|\nabla_u^\varepsilon z|^{(p-1)/2})|^2 \phi^{2p}
$$

$$
\leq C\left(\int (|\nabla_u^\varepsilon \phi|^2 + \phi^2)^p + \int |\nabla_u^\varepsilon z|^{p+1/2}\phi^{2p} + \int |X_{2u}^\varepsilon(\partial_2 u)|^p \phi^{2p} \right.
$$

$$
+ \int |f|^{2p}(|\nabla_u^\varepsilon \phi|^2 + \phi^2)\phi^{2p-2} + \int |(\nabla_u^\varepsilon)^2 u||\nabla_u^\varepsilon z|^{p-1}\phi^{2p}
$$

$$
\left. + \int |(\nabla_u^\varepsilon)^2 u|^2|\nabla_u^\varepsilon z|^{p-1}\phi^{2p} + \int |(\nabla_u^\varepsilon)^2 u||\nabla_u^\varepsilon z|^{p-1}\phi^{2p-1}|\nabla_u^\varepsilon \phi| \right).
$$

Lemma 1.52. *Let u be a smooth solution of equation (1.18) satisfying (1.19) and denote $v = \partial_2 u$. For every open set $\Omega_1 \subset\subset \Omega$ and every $p \geq 1$ there exists a positive constant C which depends on Ω_1, p, and on M in (1.19), but is independent of ε, such that*

$$
\|\nabla_u^\varepsilon u\|_{C_E^{1/2}} + \|\nabla_u^\varepsilon v\|_{L^4(\Omega_1)}^4 \leq C.
$$

Proof. We can apply Lemma 1.51 with $p = 3$ to the function $v = \partial_2 u$ and deduce that

$$
\int |(\nabla_u^\varepsilon)^2 v|^2 \phi^6 \leq C_1 + C_2\left(\int |\nabla_u^\varepsilon v|^{3+1/2}\phi^6 \right.
$$

$$
+ \int (1 + |\nabla_u^\varepsilon v| + |(\nabla_u^\varepsilon)^2 u|)^{7/5}\phi^{23/5}(|\nabla_u^\varepsilon \phi| + \phi)^{7/5} + \int |(\nabla_u^\varepsilon)^2 u||\nabla_u^\varepsilon v|^2 \phi^6
$$

$$
\left. + \int |(\nabla_u^\varepsilon)^2 u|^2|\nabla_u^\varepsilon v|^2 \phi^6 + \int |(\nabla_u^\varepsilon)^2 u||\nabla_u^\varepsilon v|^2 \phi^5|\nabla_u^\varepsilon \phi| \right).
$$

It follows that

$$
\int |(\nabla_u^\varepsilon)^2 v|^2 \phi^6 \leq \frac{C_2}{\delta} \int |(\nabla_u^\varepsilon)^2 u|^4 \phi^6 + \delta \int |\nabla_u^\varepsilon v|^4 \phi^6 + \frac{C_1}{\delta}. \tag{1.26}
$$

Analogously, if we set $z = X_{1u}^\varepsilon u$ or $z = X_{2u}^\varepsilon u$, we have

$$
\int |(\nabla_u^\varepsilon)^2 z|^2 \phi^6 \leq \frac{C_2}{\delta} \int |(\nabla_u^\varepsilon)^2 u|^4 \phi^6 + \frac{C_1}{\delta} + C_2 \int |\nabla_u^\varepsilon v|^3 \phi^6. \tag{1.27}
$$

Using Lemma 1.50, (1.26) and (1.25), we obtain immediately

$$
\int |\nabla_u^\varepsilon v|^4 \phi^6 \leq C_1 + C_2 \int |(\nabla_u^\varepsilon)^2 v|^2 \phi^6 \leq C_1 + \frac{C_2}{\delta} \int |(\nabla_u^\varepsilon)^2 u|^4 \phi^6 + \delta \int |\nabla_u^\varepsilon v|^4 \phi^6.
$$

Hence

$$\int |\nabla_u^\varepsilon v|^4 \phi^6 \leq C_1 + C_2 \int |(\nabla_u^\varepsilon)^2 u|^4 \phi^6. \tag{1.28}$$

Consequently, from the latter and (1.27) we deduce that

$$\int |(\nabla_u^\varepsilon)^2 z|^4 \phi^6 \leq C_1 + C_2 \int |(\nabla_u^\varepsilon)^2 u|^4 \phi^6. \tag{1.29}$$

Next, from the intrinsic Cacciopoli inequalities (1.28) and (1.29) we deduce an Euclidean Cacciopoli inequality. Note that

$$|\nabla_E X_{1u}^\varepsilon z| \leq |(X_{1u}^\varepsilon)^2 z| + C_2 |\partial_2 (X_{1u}^\varepsilon z)| \leq |(X_{1u}^\varepsilon)^2 z| + C_2 |v \partial_2 z| + C_2 |X_{1u}^\varepsilon \partial_2 z|$$

(since $\partial_2 z = \partial_2 X_{1u}^\varepsilon u = v^2 + X_{1u}^\varepsilon v$)

$$\leq |(\nabla_u^\varepsilon)^2 z| + C_2 |(\nabla_u^\varepsilon)^2 v| + C_2 |\nabla_u^\varepsilon v| + C_2.$$

From the latter and (1.28) and (1.29) we infer that

$$\int |\nabla_E (\nabla_u^\varepsilon) z|^2 \phi^6 \leq C_2 \left(\int |(\nabla_u^\varepsilon)^2 v|^2 \phi^6 + \int |(\nabla_u^\varepsilon)^2 z|^2 \phi^6 + 1 \right)$$

$$\leq C_2 \int |\nabla_u^\varepsilon z|^4 \phi^6 + C_1.$$

Now we can apply the standard Euclidean Sobolev inequality in $\mathbb{R}^2$ and obtain

$$\left(\int (|\nabla_u^\varepsilon z| \phi^3)^6 \right)^{1/3} \leq C_2 \int |\nabla_E (\nabla_u^\varepsilon z \phi^3)|^2 \leq C_2 \int |\nabla_u^\varepsilon z|^4 \phi^6 + C_1$$

(using Hölder's inequality)

$$\leq C_2 \left(\int (|\nabla_u^\varepsilon z| \phi^3)^6 \right)^{1/3} \left(\int_{\mathrm{supp}(\phi)} |\nabla_u^\varepsilon z|^3 \right)^{2/3} + C_1.$$

By (1.25) and the fact that $|\nabla_u^\varepsilon z| \leq |\nabla_\varepsilon^2 u|$, we already know that $|\nabla_u^\varepsilon z| \in L_{\mathrm{loc}}^3$. In fact,

$$\left(\int_{\mathrm{supp}(\phi)} |\nabla_u^\varepsilon z|^3 \right)^{2/3} \leq \left(\int_{\mathrm{supp}(\phi)} |\nabla_u^\varepsilon z|^{10/3} \right)^{3/5} |\mathrm{supp}(\phi)|^{1/15}.$$

Recall that C_2 does not depend on $|\nabla_u^\varepsilon \phi|$. If we choose the support of ϕ sufficiently small, we can assume that the integral $\int_{\mathrm{supp}(\phi)} |\nabla_u^\varepsilon z|^3$ is arbitrarily small. It follows that

$$\left(\int (|\nabla_u^\varepsilon z| \phi^3)^6 \right)^{1/3} \leq C_1,$$

and consequently, by (1.28),

$$\int |\nabla_u^\varepsilon v|^4 \phi^6 \le C_1.$$

But this implies that $|\nabla_E(\nabla_u^\varepsilon u)| \le |(\nabla_u^\varepsilon)^2 u| + |\nabla_u^\varepsilon v| + v^2 \in L_{\text{loc}}^4$. By the standard Euclidean Sobolev–Morrey inequality in $\mathbb{R}^2$, it follows that $\nabla_u^\varepsilon u \in C_E^{1/2}$. $\qquad\square$

Lemma 1.53. *Let u be a smooth solution of equation (1.18) satisfying (1.19) and denote $v = \partial_2 u$. For every open set $\Omega_1 \subset\subset \Omega$ and every $p \ge 1$ there exists a positive constant C which depends on Ω_1, p, and on M in (1.19), but is independent of ε, such that*

$$||u||_{W_\varepsilon^{2,p}(\Omega_1)} \le C.$$

Proof. We have already noted that equation (1.18) can be as well written in divergence form:

$$L_\varepsilon u = \sum_{ij} A_{ij}(\nabla_u^\varepsilon u) X_{iu}^\varepsilon X_{ju}^\varepsilon u = 0.$$

The function u satisfies uniform $C^{1,1/2}$ estimates and the coefficients $A_{ij}(\nabla_u^\varepsilon u)$ satisfy uniform $C^{1/2}$ estimates. Thus we can apply Theorem 1.40, using the fact that, for every p,

$$||\partial_2 u||_{L^p(\Omega_1)} + ||\nabla_u^\varepsilon \partial_2 u||_{L^4(\Omega_1)} \le C.$$

If follows that for every $r > 1$ there exists a constant $C > 0$ independent of ε such that $||u||_{W_\varepsilon^{2,r}(\Omega_1)} \le C$. $\qquad\square$

Using a bootstrap argument, we can now deduce the same result for derivatives of any order:

Theorem 1.54. *Let u be a smooth solution of equation (1.17) satisfying (1.19). For every open set $\Omega_1 \subset\subset \Omega$, every $p \ge 3$ and every integer $k \ge 2$, there exists a constant C which depends on p, k, Ω_1, and on M in (1.19), but is independent of ε, such that the following estimate holds:*

$$||u||_{W_\varepsilon^{k,p}(\Omega_1)} + ||\partial_2 u||_{W_\varepsilon^{k,p}(\Omega_1)} \le C.$$

Corollary 1.55. *Let u be a smooth solution of equation (1.17) satisfying (1.19). For every open set $\Omega_1 \subset\subset \Omega$, $p \ge 3$, $\alpha < 1$, and every integer $k \ge 2$, there exists a constant C which depends on p, k, Ω_1, and on M in (1.19), but is independent of ε, such that the following estimate holds:*

$$||(\nabla_u^\varepsilon)^{k+1} u||_{L^p(\Omega_1)} + ||\partial_2(\nabla_u^\varepsilon)^k u||_{L^p(\Omega_1)} + ||(\nabla_u^\varepsilon)^k u||_{C_E^\alpha(\Omega_1)} \le C.$$

1.7.4 Regularity of the viscosity minimal surface

In this subsection we turn our attention to the proof of regularity for vanishing viscosity solutions u of equation (1.17). The regularity rests on the *a priori* estimates proved in the previous subsection in the limit $\varepsilon \to 0$.

Theorem 1.56. *Let $u \in \mathrm{Lip}(\Omega)$ be a vanishing viscosity solution of (1.17). Then equation (1.17) can be represented as $X_{1u}^2 u = 0$ and is satisfied weakly in the Sobolev sense, hence pointwise a.e. in Ω, that is,*

$$\int_\Omega X_{1u} u X_{1u}^* \phi = 0 \ \ \textit{for all } \phi \in C_0^\infty(\Omega).$$

Moreover, for every $\alpha < 1$, $p > 1$, and for every natural number k,

$$(\nabla_u u)^k \in C_E^\alpha; \quad \partial_2 (\nabla_u u)^k \in W_0^{1,p}(B(R)).$$

Proof. Let (u_j) denote the sequence approximating u, as defined in Definition 1.41. For each ε_j, the function u_j is a solution of (1.18). Hence, by Corollary 1.55, the sequence $(\nabla_{u_j}^{\varepsilon_j} u_j)_j$ is bounded in C_E^α for every α. Eventually extracting a subsequence we see that it weakly converges to $(X_{1u} u, 0)$. Hence this is its limit in C_E^α norm. On the other hand, $\partial_2 u_j$ is weakly convergent to $\partial_2 u$. Hence, letting j go to ∞ in the divergence form equation, we conclude that $X_{1u}^2 u = 0$ in the weak Sobolev sense. The other part of the thesis follows from Corollary 1.55. $\quad\square$

If the weak derivatives of a function f are sufficiently regular, then they are Lie derivatives.

Proposition 1.57. *If $f \in C_{\mathrm{loc}}^\alpha(\Omega)$ for some $\alpha \in (0,1)$ and its weak derivatives satisfy $X_{1u} f \in C_{\mathrm{loc}}^\alpha(\Omega)$, $\partial_2 f \in L_{\mathrm{loc}}^p(\Omega)$ with $p > 1/\alpha$, then for all $\xi \in \Omega$ the Lie derivatives $X_{1u} f(\xi)$ exist and coincide with the weak ones.*

We are now ready to prove the result concerning the foliation:

Proof of Corollary 1.43. First note that, by Proposition 1.57, the derivatives of u are Lie derivatives. The equation $\gamma' = X_{1u} I(\gamma)$ has a unique solution, of the form

$$\gamma(x) = (x, y(x)),$$

where $y'(x) = u(x, y(x))$. In view of the regularity of u and of the previous proposition, $y''(x) = Xu(x, y(x))$ and $y'''(x) = X^2 u(x, y(x)) = 0$. This shows that γ is a polynomial of order 2 and concludes the proof. $\quad\square$

1.7.5 Foliation of minimal surfaces and completion result

Let us now present some computational results, applied to well-known images. The minimal surface which performs the completion is foliated in geodesics. This implies that each level line of the image is completed independently through an elastica, and this is compatible with the phenomenological evidence. We consider here the completion of a figure that has been only partially lifted in the rototranslation space. This example mimics the missing information due to the presence of the *macula cieca* (blind spot) that is modally completed by the human visual system, as outlined in [74]. The original image (see Fig. 1.19, top) is lifted in the rototranslation space with missing information in the center (bottom left). The lifted surface is completed iteratively until a steady state is achieved. The final surface is minimal with respect to the sub-Riemannian metric.

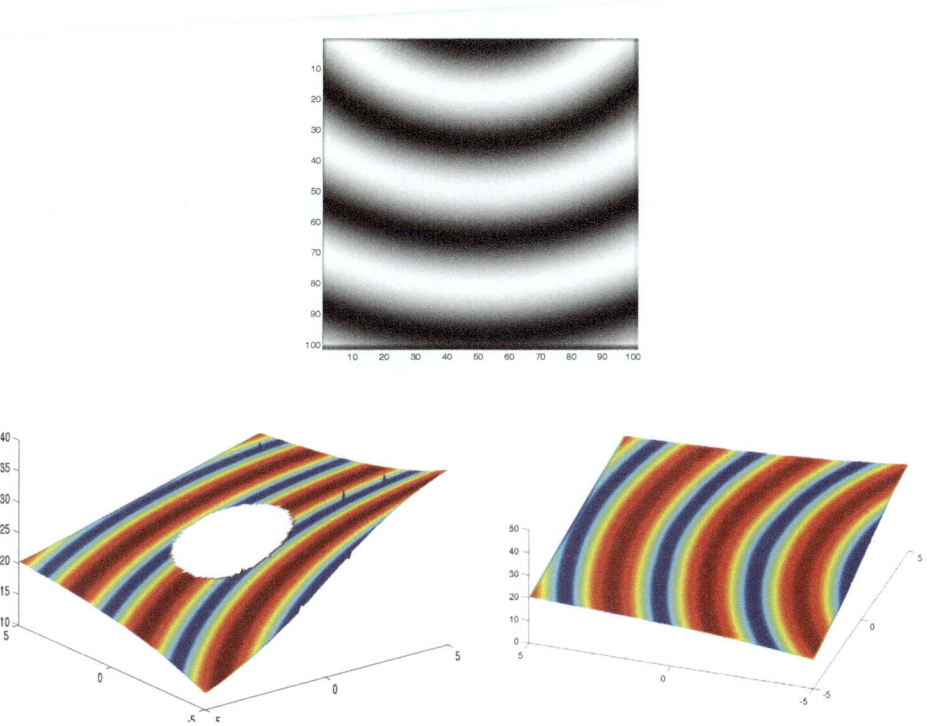

Figure 1.19: The original image (top) is lifted in the rototranslation space with missing information in the center, like in the phenomenon of *macula cieca* (bottom left). The surface is completed by the algorithm (bottom right).

Bibliography

[1] L. Ambrosio, B. Kirchheim, *Currents in metric spaces*, Acta Math. **185** (2000), 1, 1–80.

[2] L. Ambrosio, S. Masnou, *A direct variational approach to a problem arising in image reconstruction*, Interfaces Free Bound. **5** (2003), 1, 63–81.

[3] L. Ambrosio, F. Serra Cassano, D. Vittone, *Intrinsic regular hypersurfaces in Heisenberg groups*, J. Geom. Anal. **16** (2006), 2, 187–232.

[4] N. Arcozzi, F. Ferrari, *Metric normal and distance function in the Heisenberg group*, Math. Z. **256** (2007), 3, 661–684.

[5] C. Ballester, M. Bertalmio, V. Caselles, G. Sapiro, J. Verdera, *Filling-in by joint interpolation of vector fields and gray levels*, IEEE Trans. Image Process. **10** (2001), 8, 1200–1211.

[6] D. Barbieri, G. Citti, *Regularity of minimal intrinsic graphs in 3-dimensional sub-Riemannian structures of step 2*, J. Math. Pures Appl. **96** (2011), 3, 279–306.

[7] D. Barbieri, G. Citti, A. Sarti, *How uncertainty bounds the shape index of simple cells*, J. Math. Neurosci. **4** (2014), 5, 1–15.

[8] G. Barles, C. Georgelin, *A simple proof for the convergence for an approximation scheme for computing motions by mean curvature*, SIAM J. Numer. Anal. **32** (1995), 2, 484–500.

[9] V. Barone Adesi, F. Serra Cassano, D. Vittone, *The Bernstein problem for intrinsic graphs in Heisenberg groups and calibrations*, Calc. Var. Partial Differential Equations **30** (2007), 1, 17–49.

[10] G. Bellettini, R. March, *An image segmentation variational model with free discontinuities and contour curvature*, Math. Models Methods Appl. Sci. **14** (2004), 1, 1–45.

[11] G. Bellettini, M. Paolini, *Approssimazione variazionale di funzionali con curvatura*, Seminario di analisi matematica, Dip. Mat. Univ. Bologna, 1993.

[12] O. Ben-Shahar, S. Zucker, *Geometrical computations explain projection patterns of long-range horizontal connections in visual cortex*, Neural Comput. **16** (2004), 3, 445–476.

[13] G. Ben-Yosef, O. Ben-Shahar, *A tangent bundle theory for visual curve completion*, IEEE Trans. Pattern Anal. Mach. Intell. **34** (2012), 7, 1263–1280.

[14] J. Bence, B. Merriman, S. Osher, *Diffusion generated motion by mean curvature*, in: Computational Crystal Growers Workshop, J. E. Taylor (ed.), Selected Lectures in Math., Amer. Math. Soc., Providence, RI, 1992, 73–83.

[15] D. Bennequin, *Remarks on invariance in the primary visual systems of mammals*, in: Neuromathematics of Vision, G. Citti, A. Sarti (eds.), Lecture Notes in Morphogenesis, Springer, Berlin, Heidelberg, 2014.

[16] F. Bigolin, F. Serra Cassano, *Intrinsic regular graphs in Heisenberg groups vs. weak solutions of nonlinear first-order PDEs*, Adv. Calc. Var. **3** (2010), 1, 69–97.

[17] A. Bonfiglioli, E. Lanconelli, F. Uguzzoni, *Stratified Lie groups and potential theory for their sub-Laplacians*, Springer Monographs in Math., Springer, Berlin, Heidelberg, 2007.

[18] J. M. Bony, *Principe du maximum, inégalité de Harnack et unicité du problème de Cauchy pour les opérateurs elliptiques dégénérés*, Ann. Inst. Fourier (Grenoble) **19** (1969), 277–304.

[19] U. Boscain, J. Duplaix, J. P. Gauthier, F. Rossi, *Anthropomorphic image reconstruction via hypoelliptic diffusion*, SIAM J. Control Optim. **50** (2012), 3, 1309–1336.

[20] W. Bosking, Y. Zhang, B. Schofield, D. Fitzpatrick, *Orientation selectivity and the arrangement of horizontal connections in tree shrew striate cortex*, J. Neurosci. **17** (1997), 6, 2112–2127.

[21] M. Bramanti, L. Brandolini, M. Pedroni, *Basic properties of nonsmooth Hörmander's vector fields and Poincaré's inequality*, Forum Math. **25** (2013), 703–769.

[22] P. C. Bressloff, J. D. Cowan, M. Golubitsky, P. J. Thomas, M. C. Wiener, *Geometric visual hallucinations, Euclidean symmetry and the functional architecture of striate cortex*, Philos. Trans. R. Soc. Lond. B Biol. Sci. **356** (2001), 1407, 299–330.

[23] L. Capogna, G. Citti, *Generalized mean curvature flow in Carnot groups*, Comm. Partial Differential Equations **34** (2009), 937–956.

[24] L. Capogna, G. Citti, M. Manfredini, *Regularity of non-characteristic minimal graphs in the Heisenberg group* $\mathbb{H}^1$, Indiana Univ. Math. J. **5** (2009), 2115–2160.

[25] L. Capogna, G. Citti, M. Manfredini, *Smoothness of Lipschitz minimal intrinsic graphs in Heisenberg groups* $\mathbb{H}^n$, $n > 1$, J. Reine Angew. Math. **648** (2010), 75–110.

[26] L. Capogna, G. Citti, C. Senni, *Sub-Riemannian heat kernels and mean curvature flows of graphs*, J. Funct. Anal. **264** (2013), 8, 1899–1928.

[27] L. Capogna, D. Danielli, S. D. Pauls, J. T. Tyson, *An Introduction to the Heisenberg Group and the Sub-Riemannian Isoperimetric Problem*, Progress in Math., vol. 259, Birkhäuser, Basel, 2007.

[28] J.-H. Cheng, J.-F. Hwang, *Variations of generalized area functionals and p-area minimizers of bounded variation in the Heisenberg group*, Bull. Inst. Math. Acad. Sin. (N.S.) **5** (2010), 4, 369–412.

[29] J.-H. Cheng, J.-F. Hwang, *Uniqueness of p-area minimizers and integrability of a horizontal normal in the Heisenberg group*, Calc. Var. Partial Differential Equations **50** (2014), 3-4, 579–597.

[30] J.-H. Cheng, J.-F. Hwang, A. Malchiodi, P. Yang, *Minimal surfaces in pseu-dohermitian geometry*, Ann. Sc. Norm. Super. Pisa Cl. Sci. (5) **4** (2005), 1, 129–177.

[31] J.-H. Cheng, J.-F. Hwang, A. Malchiodi, P. Yang, *A Codazzi-like equation and the singular set for C^1 smooth surfaces in the Heisenberg group*, J. Reine Angew. Math. **671** (2012), 131–198.

[32] J.-H. Cheng, J.-F. Hwang, P. Yang, *Existence and uniqueness for p-area minimizers in the Heisenberg group*, Math. Ann. **337** (2007), 2, 253–293.

[33] G. Citti, E. Lanconelli, A. Montanari, *Smoothness of Lipschitz-continuous graphs with nonvanishing Levi curvature*, Acta Math. **188** (2002), 1, 87–128.

[34] G. Citti, M. Manfredini, *Blow-up in non homogeneous Lie groups and recti-fiability*, Houston J. Math. **31** (2005), 2, 333–353.

[35] G. Citti, M. Manfredini, *Implicit function theorem in Carnot-Carathéodory spaces*, Commun. Contemp. Math. **8** (2006), 5, 657–680.

[36] G. Citti, M. Manfredini, *Uniform estimates of the fundamental solution for a family of hypoelliptic operators*, Potential Anal. **25** (2006), 2, 147–164.

[37] G. Citti, A. Sarti, *A cortical based model of perceptual completion in the roto-translation space*, J. Math. Imaging Vision **24** (2006), 3, 307–326.

[38] G. Citti, A. Sarti (eds.), *Neuromathematics of Vision*, Lecture Notes in Mor-phogenesis, Springer, Berlin, Heidelberg, 2014.

[39] G. Cocci, D. Barbieri, A. Sarti, *Spatio-temporal receptive fields of cells in V1 are optimally shaped for stimulus velocity estimation*, J. Opt. Soc. Am. A **29** (2012), 1, 130–138.

[40] M. C. Crair, E. S. Ruthazer, D. C. Gillespie, M. P. Stryker, *Ocular domi-nance peaks at pinwheel center singularities of the orientation map in cat visual cortex*, J. Neurophysiol. **77** (1997), 3381–3385.

[41] D. Danielli, N. Garofalo, D.-M. Nhieu, *Sub-Riemannian calculus on hyper-surfaces in Carnot groups*, Adv. Math. **215** (2007), 1, 292–378.

[42] D. Danielli, N. Garofalo, D.-M. Nhieu, *A notable family of entire intrinsic minimal graphs in the Heisenberg group which are not perimeter minimizing*, Amer. J. Math. **130** (2008), 2, 317–339.

[43] D. Danielli, N. Garofalo, D.-M. Nhieu, S. D. Pauls, *The Bernstein problem for embedded surfaces in the Heisenberg group $\mathbb{H}^1$*, Indiana Univ. Math. J. **59** (2010), 2, 563–594.

[44] J. G. Daugman, *Uncertainty relation for resolution in space, spatial fre-quency, and orientation optimized by two-dimensional visual cortical filters*, J. Opt. Soc. Am. A **7** (1985), 2, 1160–1169.

[45] G. C. DeAngelis, I. Ohzawa, R. D. Freeman, *Receptive-field dynamics in the central visual pathways*, Trends Neurosci. **18** (1995), 451–458.

[46] E. de Giorgi, *Some remarks on Γ convergence and least square methods*, in: Composite Media and Homogenization Theory, G. Dal Maso, G. F. Dell'Antonio (eds.), Birkhäuser, Boston, 1991, 153–142.

[47] R. Duits, U. Boscain, F. Rossi, Y. Sachov, *Association fields via cuspless sub-Riemannian geodesics in* SE(2), J. Math. Imaging Vision **49** (2014), 2, 384–417.

[48] R. Duits, E. Franken, *Left-invariant parabolic evolutions on SE(2) and contour enhancement via invertible orientation scores, Part I: Linear left-invariant diffusion equations on SE(2)*, Quart. Appl. Math. **68** (2010), 2, 255–292.

[49] R. Duits, E. Franken, *Left-invariant parabolic evolutions on SE(2) and contour enhancement via invertible orientation scores, Part II: Nonlinear left-invariant diffusions on invertible orientation scores*, Quart. Appl. Math. **68** (2010), 2, 293–331.

[50] R. Duits, B. Janssen, A. Becciu, H. van Assen, *A variational approach to cardiac motion estimation based on covariant derivatives and multi-scale Helmholtz decomposition*, Quart. Appl. Math. **71** (2013), 1, 1–36.

[51] S. Esedoglu, R. March, *Segmentation with depth but without detecting junctions*, J. Math. Imaging Vision **18** (2003), 7–15.

[52] L. Evans, *Convergence of an algorithm for mean curvature motion*, Indiana Univ. Math. J. **42** (1993), 2, 553–557.

[53] H. Federer, *Geometric Measure Theory*, Grundlehren Math. Wiss., vol. 153, Springer, Berlin, Heidelberg, 1969; reprinted in: Classics in Math., 1996.

[54] D. J. Field, A. Hayes, R. F. Hess, *Contour integration by the human visual system: Evidence for a local "association field"*, Vision Res. **33** (1993), 2, 173–193.

[55] L. Florack, R. Duits, G. Jongbloed, M.-C. van Lieshout, L. Davies, *Mathematical Methods for Signal and Image Analysis and Representation*, Comput. Imaging Vision, vol. 41, Springer, Dordrecht, 2012.

[56] G. B. Folland, *Subelliptic estimates and function spaces on nilpotent Lie groups*, Ark. Mat. **13** (1975), 161–207.

[57] G. B. Folland, E. M. Stein, *Estimates for the ∂_b complex and analysis on the Heisenberg group*, Comm. Pure Appl. Math. **20** (1974), 429–522.

[58] B. Franchi, R. Serapioni, F. Serra Cassano, *Rectifiability and perimeter in the Heisenberg group*, Math. Ann. **321** (2001), 479–531.

[59] B. Franchi, R. Serapioni, F. Serra Cassano, *Rectifiability and perimeter in step 2 groups*, Mathematica Bohemica **127** (2002), 2, 219–228.

[60] B. Franchi, R. Serapioni, F. Serra Cassano, *On the structure of finite perimeter sets in step 2 Carnot groups*, J. Geom. Anal. **13** (2003), 3, 421–466.

[61] B. Franchi, R. Serapioni, F. Serra Cassano, *Regular hypersurfaces, intrinsic perimeter and implicit function theorem in Carnot groups*, Comm. Anal. Geom. **11** (2003), 5, 909–944.

[62] E. Franken, R. Duits, *Crossing-preserving coherence-enhancing diffusion on invertible orientation scores*, Int. J. Comput. Vis. **85** (2009), 3, 253–278.

[63] E. Franken, R. Duits, B. M. ter Haar Romeny, *Nonlinear diffusion on the 2D Euclidean motion group*, in: F. Sgallari, A. Murli, N. Paragios (eds.), Scale Space and Variational Methods in Computer Vision, Lecture Notes in Computer Science, vol. 4485, Springer, Berlin, Heidelberg, 2007, 461–472.

[64] M. Galli, M. Ritoré, *Area-stationary and stable surfaces of class C^1 in the sub-Riemannian Heisenberg group $\mathbb{H}^1$*, preprint, 2014, arXiv:1410.3619.

[65] N. Garofalo, D.-M. Nhieu, *Isoperimetric and Sobolev inequalities for Carnot–Carathéodory spaces and the existence of minimal surfaces*, Comm. Pure Appl. Math. **49** (1996), 10, 1081–1144.

[66] N. Garofalo, S. D. Pauls, *The Bernstein problem in the Heisenberg group*, preprint, 2002, arXiv:math/0209065.

[67] R. F. Hess, A. Hayes, D. J. Field, *Contour integration and cortical processing*, J. Physiol. (Paris) **97** (2003), 105–119.

[68] R. K. Hladky, S. D. Pauls, *Constant mean curvature surfaces in sub-Riemannian geometry*, J. Differential Geom. **79** (2008), 1, 111–139.

[69] R. K. Hladky, S. D. Pauls, *Minimal surfaces in the rototranslational group with applications to a neuro-biological image completion model*, J. Math. Imaging Vision **36** (2010), 1–27.

[70] W. C. Hoffman, *The visual cortex is a contact bundle*, Appl. Math. Comput. **32** (1989), 137–167.

[71] L. Hörmander, *Hypoelliptic second-order differential equations*, Acta Math. **119** (1967), 147–171.

[72] J. P. Jones, L. A. Palmer, *An evaluation of the two-dimensional Gabor filter model of simple receptive fields in cat striate cortex*, J. Neurophysiol. **58** (1987), 1233–1258.

[73] G. Kanizsa, *Organization in Vision*, Praeger Publishers, New York, 1979.

[74] G. Kanizsa, *Grammatica del vedere*, Il Mulino, Bologna, 1980.

[75] M. B. Karmanova, *The graphs of Lipschitz functions and minimal surfaces on Carnot groups*, Sib. Math. J. **53** (2012), 4, 672–690.

[76] B. Kirchheim, F. Serra Cassano, *Rectifiability and parametrization of intrinsic regular surfaces in the Heisenberg group*, Ann. Sc. Norm. Super. Pisa Cl. Sci. **4** (2004), 871–896.

[77] J. J. Koenderink, *The brain is a geometry engine*, Psychol. Res. **52** (1990), 122–127.

[78] B. Kunsberg, S. Zucker, *Shape-from-shading and cortical computation: a new formulation*, J. Vision **12** (2012), 9, 233.

[79] E. Lanconelli, D. Morbidelli, *On the Poincaré inequality for vector fields*, Ark. Mat. **38** (2000), 327–342.

[80] M. Manfredini, *A note on the Poincaré inequality for Lipschitz vector fields of step two*, Proc. Amer. Math. Soc. **138** (2010), 2, 567–575.

[81] M. Manfredini, *Fundamental solutions for sum of squares of vector fields operators with $C^{1,\alpha}$ coefficients*, Forum Math. **24** (2012), 5, 973–1011.

[82] S. Marcelja, *Mathematical description of the response of simple cortical cells*, J. Opt. Soc. Am. A **70** (1980), 1297–1300.

[83] S. Masnou, J. M. Morel, *Level lines based disocclusion*, Proc. 5th IEEE International Conference on Image Processing, Chicago, Illinois, October 4–7, 1998.

[84] D. McLaughlin, R. Shapley, M. Shelley, D. J. Wielaard, *A neuronal network model of macaque primary visual cortex (V1): Orientation selectivity and dynamics in the input layer $4C\alpha$*, Proc. Natl. Acad. Sci. USA **97** (2000), 14, 8087–8092.

[85] K. D. Miller, A. Kayser, N. J. Priebe, *Contrast-dependent nonlinearities arise locally in a model of contrast-invariant orientation tuning*, J. Neurophysiol. **85** (2001), 2130–2149.

[86] A. Montanari, D. Morbidelli, *Balls defined by vector fields and the Poincaré inequality*, Ann. Inst. Fourier (Grenoble) **54** (2004), 431–452.

[87] A. Montanari, D. Morbidelli, *Nonsmooth Hörmander vector fields and their control balls*, Trans. Amer. Math. Soc. **364** (2012), 5, 2339–2375.

[88] F. Montefalcone, *Hypersurfaces and variational formulas in sub-Riemannian Carnot groups*, J. Math. Pures Appl. **87** (2007), 5, 453–494.

[89] R. Montgomery, *A Tour of Subriemannian Geometries, their Geodesics and Applications*, Math. Surveys and Monographs, vol. 91, Amer. Math. Soc., Providence, RI, 2002.

[90] R. Monti, *Lipschitz approximation of H-perimeter minimizing boundaries*, Calc. Var. Partial Differential Equations **50** (2014), 1-2, 171–198.

[91] R. Monti, F. Serra Cassano, D. Vittone, *A negative answer to the Bernstein problem for intrinsic graphs in the Heisenberg group*, Boll. Unione Mat. Ital. (9) **1** (2008), 3, 709–727.

[92] D. Mumford, *Elastica and computer vision*, in: Algebraic Geometry and its Applications, C. L. Bajaj (ed.), Springer, New York, 1994, 491–506.

[93] A. Nagel, E. M. Stein, S. Wainger, *Balls and metrics defined by vector fields I: Basic properties*, Acta Math. **155** (1985), 103–147.

[94] S. B. Nelson, M. Sur, D. C. Somers, *An emergent model of orientation selectivity in cat visual cortical simple cells*, J. Neurosci. **15** (1995), 5448–5465.

[95] Y. Ni, *Sub-Riemannian constant mean curvature surfaces in the Heisenberg group as limits*, Ann. Mat. Pura Appl. **183** (2004), 4, 555–570.

[96] M. Nitzberg, D. Mumford, T. Shiota, *Filtering, Segmentation and Depth*, Springer, Berlin, Heidelberg, 1993.

[97] R. Palma-Amestoy, E. Provenzi, M. Bertalmio, V. Caselles, *A perceptually inspired variational framework for color enhancement*, IEEE Trans. Pattern Anal. Mach. Intell. **31** (2009), 3, 458–474.

[98] S. D. Pauls, *Minimal surfaces in the Heisenberg group*, Geom. Dedicata **104** (2004), 201–231.

[99] S. D. Pauls, *H-minimal graphs of low regularity in* $\mathbb{H}^1$, Comment. Math. Helv. **81** (2006), 2, 337–381.

[100] J. Petitot, *Phenomenology of perception, qualitative physics and sheaf mereology*, in: Proceedings of the 16th International Wittgenstein Symposium, Verlag Hölder-Pichler-Tempsky, Vienna, 1994, 387–408.

[101] J. Petitot, *Morphological eidetics for phenomenology of perception*, in: Naturalizing Phenomenology Issues in Contemporary Phenomenology and Cognitive Science, J. Petitot, F. J. Varela, J.-M. Roy, B. Pachoud (eds.), Stanford University Press, Stanford, 1998, 330–371.

[102] J. Petitot, Y. Tondut, *Vers une neuro-géométrie. Fibrations corticales, structures de contact et contours subjectifs modaux*, Mathématiques, Informatique et Sciences Humaines, vol. 145, EHESS, Paris, 1999, 5–101.

[103] N. J. Priebe, K. D. Miller, T. W. Troyer, A. E. Krukowsky, *Contrast-invariant orientation tuning in cat visual cortex: thalamocortical input tuning and correlation-based intracortical connectivity*, J. Neurosci. **18** (1998), 5908–5927.

[104] M. Ritoré, *Examples of area-minimizing surfaces in the sub-Riemannian Heisenberg group* $\mathbb{H}^1$ *with low regularity*, Calc. Var. Partial Differential Equations **34** (2009), 2, 179–192.

[105] M. Ritoré, C. Rosales, *Rotationally invariant hypersurfaces with constant mean curvature in the Heisenberg group* $\mathbb{H}^n$, J. Geom. Anal. **16** (2006), 4, 703–720.

[106] M. Ritoré, C. Rosales, *Area-stationary surfaces in the Heisenberg group* $\mathbb{H}^1$, Adv. Math. **219** (2008), 633–671.

[107] L. Rothschild, E. M. Stein, *Hypoelliptic differential operators and nilpotent Lie groups*, Acta Math. **137** (1977), 247–320.

[108] G. Sanguinetti, G. Citti, A. Sarti, *Implementation of a model for perceptual completion in* $\mathbb{R}^2 \times S^1$, in: Computer Vision and Computer Graphics, Theory and Applications, vol. 24, Springer, Berlin, Heidelberg, 2009, 188–201.

[109] A. Sarti, *A cortical based model of perceptual completion in the roto-translation space: Part I*, in: Proceedings of the Workshop on Second Order Subelliptic Equations and Applications, Cortona, June 15–21, 2003.

[110] A. Sarti, G. Citti, *The constitution of perceptual units in the functional architecture of* V1, to appear in J. Comput. Neurosci., arXiv:1406.0289.

[111] A. Sarti, G. Citti, J. Petitot, *The symplectic structure of the primary visual cortex*, Biol. Cybernet. **98** (2008), 1, 33–48.

[112] A. Sarti, G. Citti, J. Petitot, *Functional geometry of the horizontal connectivity in the primary visual cortex*, J. Physiol. (Paris) **103** (2009), 1-2, 37–45.

[113] A. Sarti, R. Malladi, J. A. Sethian, *Subjective surfaces: A method for completing missing boundaries*, Proc. Natl. Acad. Sci. USA **97** (2000), 12, 6258–6263.

[114] F. Serra Cassano, D. Vittone, *Graphs of bounded variation, existence and local boundedness of non-parametric minimal surfaces in the Heisenberg group*, Adv. Calc. Var. **7** (2014), 4, 409–492.

[115] N. Shcherbakova, *Minimal surfaces in sub-Riemannian manifolds and structure of their singular sets in the* $(2,3)$ *case*, ESAIM Control Optim. Calc. Var. **15** (2009), 4, 839–862.

[116] D. Ts'o, C. D. Gilbert, T. N. Diesel, *Relationship between horizontal interactions and functional architecture in cat striate cortex as revealed by cross-correlation analysis*, J. Neurosci. **6** (1986), 4, 1160–1170.

[117] S. W. Zucker, *The curve indicator random field: curve organization via edge correlation*, in: Perceptual Organization for Artificial Vision Systems, K. L. Boyer, S. Sarkar (eds.), Kluwer Academic, Boston, 2000, 265–288.

[118] J. Zweck, L. R. Williams, *Euclidean group invariant computation of stochastic completion fields using shiftable-twistable functions*, J. Math. Imaging Vis. **21** (2004), 2, 135–154.

Chapter 2

Multilinear Calderón–Zygmund Singular Integrals

Loukas Grafakos[1]

2.1 Introduction

It is quite common for linear operators to depend on several functions of which only one is thought of as the main variable and the remaining ones are usually treated as parameters. Examples of such operators are ubiquitous in harmonic analysis: multiplier operators, homogeneous singular integrals associated with functions Ω on the sphere, Littlewood–Paley operators, Calderón commutators, and the Cauchy integral along Lipschitz curves. Treating the additional functions that arise in these operators as frozen parameters often provides limited results that could be thought analogous to those that one obtains by studying calculus of functions of several variables by freezing variables. In this article, we advocate a more flexible point of view in the study of linear operators, analogous to that employed in pure multivariable calculus. Unfreezing the additional functions and treating them as input variables provides a more robust approach that often yields sharper results in terms of regularity of the input functions.

[1]This chapter contains the material covered in a minicourse given by the author at the Centre de Recerca Matemàtica in Barcelona during the period May 4–8, 2009. The course was an expanded version of a series of three lectures delivered by the author two weeks earlier (April 23–25) at the New Mexico Analysis Seminar held at the University of New Mexico in Albuquerque. The author would like to thank Joan Mateu and Joan Orobitg for coordinating two special minicourses at the Centre de Recerca Matemàtica on Multilinear Harmonic Analysis and Weights presented by the author and Professor Carlos Pérez in May 2009. These minicourses were part of a special research program for the academic year 2008–2009 entitled *Harmonic Analysis, Geometric Measure Theory, and Quasiconformal Mappings*, coordinated by Xavier Tolsa and Joan Verdera. The author would also like to thank the Centre de Recerca Matemàtica for providing an inspiring environment for research to the participants during these workshops. This research was partially supported by the NSF under grant DMS 0900946.

We illustrate the power of this idea with a concrete example concerning the Hörmander–Mihlin multiplier theorem [23], [34]. This says that for $\gamma > n/2$ there is a constant $C_{p,n,\gamma}$ such that

$$\left\|\left(\widehat{f}\sigma\right)^{\vee}\right\|_{L^p(\mathbf{R}^n)} \tag{2.1}$$

$$\leq C_{p,n,\gamma}\|f\|_{L^p(\mathbf{R}^n)}\left[\|\sigma\|_{L^\infty(\mathbf{R}^n)} + \sup_{k\in\mathbf{Z}}\left\|\varphi(\xi)\sigma(2^k\xi)\right\|_{L^2_\gamma(\mathbf{R}^n,d\xi)}\right]$$

where $1 < p < \infty$, while for $p = 1$ the inequality is still valid when the L^p norm on the left is replaced by $L^{1,\infty}$. Here φ is a smooth function supported in the annulus $1/2 < |\xi| < 2$ which is nonvanishing in the smaller annulus $1/\sqrt{2} < |\xi| < \sqrt{2}$, and $L^r_\gamma(\mathbf{R}^n)$ denotes the Sobolev space of functions on $\mathbf{R}^n$ with norm

$$\|\sigma\|_{L^r_\gamma(\mathbf{R}^n)} = \left(\int_{\mathbf{R}^n}\left|\left(\widehat{\sigma}(\xi)(1+|\xi|^2)^{\gamma/2}\right)^{\vee}(x)\right|^r dx\right)^{1/r}.$$

It makes sense to view the multiplier operator $f \mapsto \left(\widehat{f}\sigma\right)^{\vee}$ as a bilinear operator acting on f and σ, that is,

$$(f,\sigma) \longmapsto B(f,\sigma) = \left(\widehat{f}\sigma\right)^{\vee}.$$

Then, when $\beta > n/q$ and for q near infinity, we have

$$\left\|B(f,\sigma)\right\|_{L^2} \leq \|\sigma\|_{L^\infty}\|f\|_{L^2} \leq C_{n,q,\beta}\|\sigma\|_{L^q_\beta}\|f\|_{L^2}, \tag{2.2}$$

where the last estimate follows by the Sobolev embedding theorem. To make the example more transparent, let us assume that σ is supported in a compact set that does not contain the origin. In view of the assumption on the support of σ and the Sobolev embedding theorem, the estimate (2.1) with $p = 1$ reduces to

$$\left\|B(f,\sigma)\right\|_{L^{1,\infty}(\mathbf{R}^n)} \leq C_{n,\gamma}\|\sigma\|_{L^2_\gamma(\mathbf{R}^n)}\|f\|_{L^1(\mathbf{R}^n)} \tag{2.3}$$

whenever $\gamma > n/2$. Interpolating bilinearly between (2.2) and (2.3) in the complex way, we deduce for $1 \leq p < 2$ the inequality

$$\left\|B(f,\sigma)\right\|_{L^{p,p'}(\mathbf{R}^n)} \leq C_{n,p,\delta}\|\sigma\|_{L^r_\delta(\mathbf{R}^n)}\|f\|_{L^p(\mathbf{R}^n)}, \tag{2.4}$$

where

$$\frac{1}{r} = \frac{1}{p} - \frac{1}{2}$$

and

$$\delta > \frac{n}{r}.$$

Further use of interpolation provides an improved version of (2.4) in which $L^{p,p'}$ is replaced by the smaller space L^p whenever $1 < p < 2$. The advantage of (2.4) versus (2.1) is that, for p near 2, the number of derivatives required of σ in (2.4) is proportional to the distance of p from 2, while in (2.1) this number of derivatives remains constant for all p. This improvement is a result of bilinear interpolation and of the approach of treating the linear multiplier operator $f \mapsto (\widehat{f}\sigma)^{\vee}$ as a bilinear operator of both functions f and σ.

Having made the point that important information can be extracted by unfreezing parameters and treating them as variables, in the sequel we pursue this idea in a more systematic way. The purpose of these lectures is to present certain fundamental results concerning linear operators of several variables, henceforth called multilinear, that indicate some of the unique challenges that appear in their study, albeit the great similarities they share with their linear counterparts. For the purpose of clarity in the presentation, we only discuss these results in the bilinear case. The proofs that we provide may not contain all necessary details, but references are provided.

If a bilinear operator T commutes with translations, in the sense that

$$T(f, g)(x + t) = T(f(\cdot + t), g(\cdot + t))(x) \tag{2.5}$$

for all $t, x \in \mathbf{R}^n$, then it incorporates a certain amount of homogeneity. Indeed, if it maps $L^{p_1} \times L^{p_2}$ to L^p, then one must necessarily have $1/p_1 + 1/p_2 \geq 1/p$; this was proved in [22] for compactly supported kernels but extended for general kernels in recent work [12]. The situation where $1/p_1 + 1/p_2 = 1/p$ will be referred to as the singular integral case. Bilinear operators that commute with translations as in (2.5) are exactly the bilinear multiplier operators that have the form

$$T(f_1, f_2)(x) = \int_{\mathbf{R}^n} \int_{\mathbf{R}^n} \sigma(\xi_1, \xi_2)\, \widehat{f_1}(\xi_1)\, \widehat{f_2}(\xi_2)\, e^{2\pi i x \cdot (\xi_1 + \xi_2)} \, d\xi_1 \, d\xi_2$$

for some bounded function σ. The situation $1/p_1 + 1/p_2 > 1/p$ may also arise. For instance, the fractional integrals

$$J_\alpha(f, g)(x) = \int_{\mathbf{R}^n} f(x - t)\, g(x + t)\, |t|^{\alpha - n} \, dt$$

and

$$J'_\alpha(f, g)(x) = \int_{\mathbf{R}^n} \int_{\mathbf{R}^n} f(x - t)\, g(x - s)\, (|t| + |s|)^{\alpha - 2n} \, dt \, ds$$

map $L^{p_1} \times L^{p_2}$ to L^p whenever $1/p_1 + 1/p_2 = 1/p + \alpha/n$. The estimate for J'_α is trivial as this operator is pointwise controlled by the product of two linear fractional integrals, but the corresponding estimate for J_α requires more work; see [15], [27].

Endpoint estimates for linear singular integrals are usually estimates of the form $L^1 \to L^1$ or $L^1 \to L^{1,\infty}$. The analogous bilinear estimates are of the form $L^1 \times L^1 \to L^{1/2,\infty}$. Although one expects some similarities with the linear case, there exist some differences as well. For example, if a linear translation-invariant operator has a positive kernel and it maps L^1 to $L^{1,\infty}$, then it must have an integrable kernel and thus it actually maps L^1 to L^1. In the bilinear case, it is still true that if a bilinear translation-invariant operator has a positive kernel and maps $L^1 \times L^1$ to $L^{1/2,\infty}$, then it must have an integrable kernel. However, having an integrable positive kernel does not necessarily imply that the corresponding operator maps $L^1 \times L^1$ to $L^{1/2}$. Results of this type have been studied in [20].

We end this section by discussing certain examples of bilinear operators.

Example 2.1. The "identity operator" in the bilinear setting is the product operator

$$B_1(f,g)(x) = f(x)\,g(x).$$

In view of Hölder's inequality, B_1 maps $L^p \times L^q \to L^r$ whenever $1/p + 1/q = 1/r$.

Example 2.2. The action of a linear operator L on the product fg gives rise to a more general "degenerate bilinear operator"

$$B_2(f,g)(x) = L(f\,g)(x),$$

that still maps $L^p \times L^q \to L^r$ whenever $1/p + 1/q = 1/r$, provided L is a bounded operator on L^r.

Example 2.3. This example captures all interesting bilinear operators. Let $\widetilde{L}$ be a linear operator acting on functions defined on $\mathbf{R}^{2n}$. Then, for functions f, g on $\mathbf{R}^n$, we consider the tensor function $(f \otimes g)(x,y) = f(x)g(y)$ and define

$$B_3(f,g)(x) = \widetilde{L}(f \otimes g)(x,x).$$

In particular, $\widetilde{L}$ could be a singular integral acting on functions on $\mathbf{R}^{2n}$. Boundedness of B_3 from $L^p(\mathbf{R}^n) \times L^q(\mathbf{R}^n) \to L^r(\mathbf{R}^n)$ is a delicate issue and will be investigated in this article for certain classes of linear operators $\widetilde{L}$.

2.2   Bilinear Calderón–Zygmund operators

It is appropriate to embark on our study with the class of operators that extend the concept of Calderón–Zygmund operators in the multilinear setting. These operators have kernels that satisfy standard estimates and possess boundedness properties analogous to those of the classical linear ones. This class of operators has been previously studied by Coifman and Meyer [6], [7], [8], [9], [33], assuming sufficient smoothness on their symbols and kernels. This area of research is still quite active. Recent developments include the introduction of a new class of multiple weights appropriate for the boundedness of these operators on weighted Lebesgue spaces; see [31].

We will be working on n-dimensional space $\mathbf{R}^n$. We denote by $\mathcal{S}(\mathbf{R}^n)$ the space of all Schwartz functions on $\mathbf{R}^n$ and by $\mathcal{S}'(\mathbf{R}^n)$ its dual space, the set of all tempered distributions on $\mathbf{R}^n$. We use the following definition for the Fourier transform in n-dimensional Euclidean space:

$$\widehat{f}(\xi) = \int_{\mathbf{R}^n} f(x) \, e^{-2\pi i x \cdot \xi} \, dx,$$

while $f^{\vee}(\xi) = \widehat{f}(-\xi)$ denotes the inverse Fourier transform. A bilinear operator $T \colon \mathcal{S}(\mathbf{R}^n) \times \mathcal{S}(\mathbf{R}^n) \to \mathcal{S}'(\mathbf{R}^n)$ is linear in every entry and consequently has two formal transposes. The first transpose T^{*1} of T is defined via

$$\langle T^{*1}(f_1, f_2), \, h \rangle = \langle T(h, f_2), \, f_1 \rangle,$$

for all f_1, f_2, h in $\mathcal{S}(\mathbf{R}^n)$. Analogously one defines T^{*2} and we also set $T^{*0} = T$.

Let $K(x, y_1, y_2)$ be a locally integrable function defined away from the *diagonal* $x = y_1 = y_2$ in $(\mathbf{R}^n)^3$, which satisfies the *size estimate*

$$|K(x, y_1, y_2)| \leq \frac{A}{(|x - y_1| + |x - y_2|)^{2n}} \tag{2.6}$$

for some $A > 0$ and all $(x, y_1, y_2) \in (\mathbf{R}^n)^3$ with $x \neq y_j$ for some j. Furthermore, assume that for some $\varepsilon > 0$ we have the smoothness estimates

$$|K(x, y_1, y_2) - K(x', y_1, y_2)| \leq \frac{A \, |x - x'|^{\varepsilon}}{(|x - y_1| + \ |x - y_2|)^{2n+\varepsilon}} \tag{2.7}$$

whenever $|x - x'| \leq \frac{1}{2} \max \left(|x - y_1|, |x - y_2| \right)$, and also that

$$|K(x, y_1, y_2) - K(x, y_1', y_2)| \leq \frac{A \, |y_j - y_j'|^{\varepsilon}}{(|x - y_1| + \ |x - y_2|)^{2n+\varepsilon}} \tag{2.8}$$

whenever $|y_1 - y_1'| \leq \frac{1}{2} \max \left(|x - y_1|, |x - y_2| \right)$, as well as a similar estimate with the roles of y_1 and y_2 reversed. Kernels satisfying these conditions are called *bilinear Calderón–Zygmund kernels* and are denoted by 2-CZK(A, ε). A bilinear operator T is said to be *associated* with K if

$$T(f_1, f_2)(x) = \int_{(\mathbf{R}^n)^2} K(x, y_1, y_2) \, f_1(y_1) \, f_2(y_2) \, dy_1 \, dy_2 \tag{2.9}$$

whenever f_1, f_2 are smooth functions with compact support and x does not lie in the intersection of the supports of f_1 and f_2.

Certain homogeneous distributions of order $-2n$ are examples of kernels in the class 2-CZK(A, ε). For this reason, boundedness properties of operators T with kernels in 2-CZK(A, ε) from a product $L^{p_1} \times L^{p_2}$ into another L^p space can only hold when

$$\frac{1}{p_1} + \frac{1}{p_2} = \frac{1}{p},$$

as dictated by homogeneity. If such boundedness holds for a certain triple of Lebesgue spaces, then the corresponding operator is called bilinear Calderón–Zygmund.

The first main result concerning these operators is the bilinear extension of the classical Calderón–Zygmund [3]; the linear result states that, if an operator with smooth enough kernel is bounded on a certain L^r space, then it is of weak type $(1,1)$ and is also bounded on all L^p spaces for $1 < p < \infty$. A multilinear version of this theorem has been obtained by Grafakos and Torres [22] for operators with kernels in the class 2-CZK(A, ε). A special case of this result was also obtained by Kenig and Stein [27]; all approaches build on previous work by Coifman and Meyer [6].

Theorem 2.4. *Let T be a bilinear operator with kernel K in 2-CZK(A, ε). Assume that, for some $1 \leq q_1, q_2 \leq \infty$ and some $0 < q < \infty$ with*

$$\frac{1}{q_1} + \frac{1}{q_2} = \frac{1}{q},$$

T maps $L^{q_1} \times L^{q_2}$ to $L^{q,\infty}$. Then T can be extended to a bounded operator from $L^1 \times L^1$ into $L^{1/2,\infty}$. Moreover, for some constant C_n (that depends only on the parameters indicated) we have that

$$\|T\|_{L^1 \times L^1 \to L^{1/2,\infty}} \leq C_n \big(A + \|T\|_{L^{q_1} \times L^{q_2} \to L^{q,\infty}} \big). \tag{2.10}$$

Proof. Set $B = \|T\|_{L^{q_1} \times L^{q_2} \to L^{q,\infty}}$. Fix an $\alpha > 0$ and consider functions $f_j \in L^1$ for $1 \leq j \leq 2$. Without loss of generality we may assume that $\|f_1\|_{L^1} = \|f_2\|_{L^1} = 1$. Setting $E_\alpha = \{x : |T(f_1, f_2)(x)| > \alpha\}$, we need to show that for some constant $C = C_n$ we have

$$|E_\alpha| \leq C(A + B)^{1/2} \alpha^{-1/2}. \tag{2.11}$$

(Once (2.11) has been established for f_j's with norm one, the general case follows immediately by scaling.) Let γ be a positive real number to be determined later. For each $j = 1, 2$, apply the Calderón–Zygmund decomposition to the function f_j at height $(\alpha\gamma)^{1/2}$ to obtain 'good' and 'bad' functions g_j and b_j, and families of disjoint cubes $\{Q_{j,k}\}_k$, such that $f_j = g_j + b_j$ and $b_j = \sum_k b_{j,k}$, where

$$\text{support}(b_{j,k}) \subset Q_{j,k},$$

$$\int_{Q_{j,k}} b_{j,k}(x)\,dx = 0, \qquad \int_{Q_{j,k}} |b_{j,k}(x)|\,dx \leq C(\alpha\gamma)^{1/2}|Q_{j,k}|,$$

$$\Big| \cup_k Q_{j,k} \Big| \leq C(\alpha\gamma)^{-1/2}, \qquad \|b_j\|_{L^1} \leq C, \qquad \|g_j\|_{L^s} \leq C(\alpha\gamma)^{1/2s'},$$

for any $1 \leq s \leq \infty$ (here s' is the dual exponent of s). Now let

$$E_1 = \{x : |T(g_1, g_2)(x)| > \alpha/4\},$$
$$E_2 = \{x : |T(b_1, g_2)(x)| > \alpha/4\},$$

$$E_3 = \{x \ : \ |T(g_1, b_2)(x)| > \alpha/4\},$$
$$E_4 = \{x \ : \ |T(b_1, b_2)(x)| > \alpha/4\}.$$

Since $|\{x \ : \ |T(f_1, f_2)(x)| > \alpha\}| \leq \sum_{s=1}^{4} |E_s|$, it will suffice to prove estimate (2.11) for each E_s. Chebychev's inequality and $L^{q_1} \times L^{q_2} \to L^{q,\infty}$ boundedness give

$$|E_1| \leq \frac{(4B)^q}{\alpha^q} \|g_1\|_{L^{q_1}}^q \|g_2\|_{L^{q_2}}^q \leq \frac{CB^q}{\alpha^q} \prod_{j=1}^{2} (\alpha\gamma)^{q/(2q_j')}$$

$$= \frac{C'B^q}{\alpha^q} (\alpha\gamma)^{(2-\frac{1}{q})\frac{q}{2}} = C'B^q \alpha^{-\frac{1}{2}} \gamma^{q-\frac{1}{2}}. \tag{2.12}$$

We now show that

$$|E_s| \leq C\alpha^{-1/2}\gamma^{1/2}. \tag{2.13}$$

Let $l(Q)$ denote the side-length of a cube Q and let Q^* be a certain expansion of it with the same center. (This expansion only depends on the dimension.) Fix an $x \notin \cup_{j=1}^{2} \cup_{k_j} (Q_{j,k_j})^*$. Also fix the cube Q_{1,k_1} and let c_{1,k_1} be its center. For fixed $y_2 \in \mathbf{R}^n$, the mean value property of the function b_{1,k_1} gives

$$\left| \int_{Q_{1,k_1}} K(x, y_1, y_2) \, b_{1,k_1}(y_1) \, dy_1 \right|$$

$$= \left| \int_{Q_{1,k_1}} \left(K(x, y_1, y_2) - K(x, c_{1,k_1}, y_2) \right) b_{1,k_1}(y_1) \, dy_1 \right|$$

$$\leq \int_{Q_{1,k_1}} |b_{1,k_1}(y_1)| \frac{A \, |y_1 - c_{1,k_1}|^\varepsilon}{(|x - y_1| + |x - y_2|)^{2n+\varepsilon}} \, dy_1$$

$$\leq \int_{Q_{1,k_1}} |b_{1,k_1}(y_1)| \frac{C \, A \, l(Q_{1,k_1})^\varepsilon}{(|x - y_1| + |x - y_2|)^{2n+\varepsilon}} \, dy_1,$$

where in the previous to last inequality we used that

$$|y_1 - c_{1,k_1}| \leq c_n \, l(Q_{1,k_1}) \leq \frac{1}{2}|x - y_1| \leq \frac{1}{2} \max \left(|x - y_1|, |x - y_2| \right).$$

Multiplying the derived inequality

$$\left| \int_{Q_{1,k_1}} K(x, y_1, y_2) \, b_{1,k_1}(y_1) \, dy_1 \right| \leq \int_{Q_{1,k_1}} \frac{C \, A \, |b_{1,k_1}(y_1)| \, l(Q_{1,k_1})^\varepsilon}{(|x - y_1| + |x - y_2|)^{2n+\varepsilon}} \, dy_1$$

by $|g_2(y_2)|$ and integrating over y_2, we obtain the estimate

$$\int_{\mathbf{R}^n} |g_2(y_2)| \left| \int_{Q_{1,k_1}} K(x, y_1, y_2)\, b_{1,k_1}(y_1)\, dy_1 \right| dy_2$$

$$\leq \|g_2\|_{L^\infty} \int_{Q_{1,k_1}} |b_{1,k_1}(y_1)| \frac{A\, C\, l(Q_{1,k_1})^\varepsilon}{(|x - y_1| + |x - y_2|)^{2n-n+\varepsilon}}\, dy_1 \qquad (2.14)$$

$$\leq C\, A\, \|g_2\|_{L^\infty} \|b_{1,k_1}\|_{L^1} \frac{l(Q_{1,k_1})^\varepsilon}{\left(l(Q_{1,k_1}) + |x - c_{1,k_1}| \right)^{n+\varepsilon}}.$$

The last inequality is due to the fact that for $x \notin \cup_{j=1}^2 \cup_{k_j} (Q_{j,k_j})^*$ and $y_j \in Q_{j,k_j}$ we have that $|x - y_j| \approx l(Q_{j,k_j}) + |x - c_{j,k_j}|$. It is now a simple consequence of (2.14) that for $x \notin \cup_{j=1}^2 \cup_{k_j} (Q_{j,k_j})^*$ we have

$$|T(b_1, g_2)(x)| \leq C' A\, (\alpha\gamma)^{\frac{1}{2}} \left(\sum_{k_1} \frac{(\alpha\gamma)^{1/2}\, l(Q_{1,k_1})^{n+\varepsilon}}{(l(Q_{1,k_1}) + |x - c_{1,k_1}|)^{n+\varepsilon}} \right) = C'' A\, \alpha\, \gamma M_\varepsilon(x),$$

$$(2.15)$$

where

$$M_\varepsilon(x) = \sum_k \frac{l(Q_{1,k})^{n+\varepsilon}}{(l(Q_{1,k}) + |x - c_{1,k}|)^{n+\varepsilon}}$$

is the Marcinkiewicz function associated with the family of cubes $\{Q_{1,k}\}_k$. It is a known fact [41] that for some constant C there is an estimate

$$\int_{\mathbf{R}^n} M_\varepsilon(x)\, dx \leq C |\cup_k Q_{1,k}| \leq C'(\alpha\gamma)^{-1/2}. \qquad (2.16)$$

Using (2.15) and (2.16), an L^1 estimate outside $\cup_{j=1}^2 \cup_{k_j} (Q_{j,k_j})^*$ gives

$$|\{x \notin \cup_{j=1}^2 \cup_{k_j} (Q_{j,k_j})^* : |T(b_1, g_2)(x)| > \alpha/4\}| \leq C\, \alpha^{-1/2} A \gamma^{1/2}. \qquad (2.17)$$

This estimate, in conjunction with

$$|\cup_{j=1}^2 \cup_{k_j} (Q_{j,k_j})^*| \leq C(\alpha\gamma)^{-1/2},$$

yields the required inequality (2.13).

We have now proved (2.13) for $\gamma > 0$. Plugging in the value of $\gamma = (A+B)^{-1}$ in both (2.12) and (2.13) gives the required estimate (2.11) for $|E_2|$. The estimate for $|E_3|$ is symmetric, while the analogous estimate for $|E_4|$ requires a variation of the argument for $|E_2|$. Since two bad functions show up in this estimate, a double sum over pairs of cubes appears and one has to use cancellation with respect to the smallest cube in the pair. Then the length of the smaller cube in the numerator is controlled by the square root of the length of the smaller cube times the square root of the length of the larger cube. At the end, the term $|T(b_1, b_2)|$ is pointwise controlled by a product of Marcinkiewicz functions outside the union $\cup_{j=1}^{2} \cup_{k_j} (Q_{j,k_j})^*$ and one uses an $L^{1/2}$ estimate over this set (instead of an L^1 estimate) in conjunction with Hölder's inequality. The previous choice of $\gamma = (A + B)^{-1}$ yields the required estimate for $|E_4|$. $\qquad\square$

Example 2.5. Let R_1 be the *bilinear Riesz transform in the first variable*

$$R_1(f_1, f_2)(x) = \text{p.v.} \int_{\mathbf{R}} \int_{\mathbf{R}} \frac{x - y_1}{|(x - y_1, x - y_2)|^3} f_1(y_1) f_2(y_2) \, dy_1 \, dy_2.$$

We will show later that this operator maps $L^{p_1}(\mathbf{R}) \times L^{p_2}(\mathbf{R})$ to $L^p(\mathbf{R})$ for $1/p_1 + 1/p_2 = 1/p$, $1 < p_1, p_2 < \infty$, and $1/2 < p < \infty$. Thus by Theorem 2.4 it also maps $L^1 \times L^1$ to $L^{1/2,\infty}$. However, it does not map $L^1 \times L^1$ to any Lorentz space $L^{1/2,q}$ for $q < \infty$. In fact, letting $f_1 = f_2 = \chi_{[0,1]}$, an easy computation shows that $R_1(f_1, f_2)(x)$ behaves at infinity like $|x|^{-2}$. This fact indicates that in Theorem 2.4 the space $L^{1/2,\infty}$ is best possible and cannot be replaced by any smaller space. In particular, it cannot be replaced by $L^{1/2}$.

2.3 Endpoint estimates and interpolation for bilinear Calderón–Zygmund operators

The real bilinear interpolation is significantly more complicated than the linear one. Early versions appeared in the work of Janson [25] and Strichartz [43]. In this exposition we will use a version of real bilinear interpolation appearing in [15]. This makes use of the notion of *bilinear restricted weak type (p, q, r) estimates*. These are estimates of the form

$$\lambda \big| \{ x : |T(\chi_A, \chi_B)(x)| > \lambda \} \big|^{1/r} \leq M \, |A|^{1/p} |B|^{1/q}$$

and have a wonderful interpolation property: if an operator T satisfies restricted weak type (p_0, q_0, r_0) and (p_1, q_1, r_1) estimates with constants M_0 and M_1, respectively, then it also satisfies a restricted weak type (p, q, r) estimate with constant $M_0^{1-\theta} M_1^{\theta}$, where

$$\left(\frac{1}{p}, \frac{1}{q}, \frac{1}{r} \right) = (1 - \theta) \left(\frac{1}{p_0}, \frac{1}{q_0}, \frac{1}{r_0} \right) + \theta \left(\frac{1}{p_1}, \frac{1}{q_1}, \frac{1}{r_1} \right).$$

We will make use of the following bilinear interpolation result; for a proof see [15].

Theorem 2.6. *Let $0 < p_{ij}, p_i \leq \infty$, $i = 1, 2, 3$, $j = 1, 2$, and suppose that the points $(1/p_{11}, 1/p_{12})$, $(1/p_{21}, 1/p_{22})$, $(1/p_{31}, 1/p_{32})$ are the vertices of a nontrivial triangle in $\mathbf{R}^2$. Let $(1/q_1, 1/q_2)$ be in the interior of this triangle (i.e., a linear convex combination of the three vertices of the triangle) and suppose that $(1/q_1, 1/q_2, 1/q)$ is the same linear combination of the points $(1/p_{11}, 1/p_{12}, 1/p_1)$, $(1/p_{21}, 1/p_{22}, 1/p_2)$, and $(1/p_{31}, 1/p_{32}, 1/p_3)$. Suppose that a bilinear operator T satisfies restricted weak type (p_{i1}, p_{i2}, p_i) estimates for $i = 1, 2, 3$. Then T has a bounded extension from $L^{q_1} \times L^{q_2}$ to L^q whenever $1/q \leq 1/q_1 + 1/q_2$.*

There is an interpolation theorem saying that if a linear operator (that satisfies a mild assumption) and its transpose are of restricted weak type $(1, 1)$, then the operator is L^2 bounded. We begin this section by proving a bilinear analogue of this result. For a more detailed version of the result below, see [21].

Theorem 2.7. *Let* $1 < p_1, p_2 < \infty$ *be such that* $1/p_1 + 1/p_2 = 1/p < 1$. *Suppose that a bilinear operator has the property that*

$$\sup_{A_0, A_1, A_2} |A_0|^{-1/p'} |A_1|^{-1/p_1} |A_2|^{-1/p_2} \left| \int_{A_0} T(\chi_{A_1}, \chi_{A_2}) \, dx \right| < \infty, \qquad (2.18)$$

where the supremum is taken over all subsets A_0, A_1, A_2 *of finite measure. Also suppose that* T, T^{*1}, *and* T^{*2} *are of restricted weak type* $(1, 1, 1/2)$; *this means that these operators map* $L^1 \times L^1$ *to* $L^{1/2, \infty}$ *when restricted to characteristic functions with constants* B_0, B_1, B_2, *respectively. Then there is a constant* C_{p_1, p_2} *such that* T *maps* $L^{p_1, 1} \times L^{p_2, 1}$ *to* $L^{p, \infty}$ *when restricted to characteristic functions with norm at most*

$$C_{p_1, p_2} B_0^{1/(2p)} B_1^{1/(2p_1')} B_2^{1/(2p_2')}.$$

Proof. Let M be the supremum in (2.18). We consider the following cases.

Case 1: Suppose that

$$\frac{|A_0|}{\sqrt{B_0}} \ge \max\left(\frac{|A_1|}{\sqrt{B_1}}, \frac{|A_2|}{\sqrt{B_2}} \right).$$

Since T maps $L^1 \times L^1$ to weak $L^{1/2}$ when restricted to characteristic functions, there exists a subset A_0' of A_0 of measure $|A_0'| \ge \frac{1}{2}|A_0|$ such that

$$\left| \int_{A_0'} T(\chi_{A_1}, \chi_{A_2}) \, dx \right| \le C B_0 |A_1| |A_2| |A_0|^{1 - \frac{1}{1/2}}$$

for some constant C. Then

$$\left| \int_{A_0} T(\chi_{A_1}, \chi_{A_2}) \, dx \right| \le \left| \int_{A_0'} T(\chi_{A_1}, \chi_{A_2}) \, dx \right| + \left| \int_{A_0 \setminus A_0'} T(\chi_{A_1}, \chi_{A_2}) \, dx \right|$$

$$\le C B_0 |A_1| |A_2| |A_0|^{-1} + M |A_1|^{1/p_1} |A_2|^{1/p_2} \left(\frac{1}{2} |A_0| \right)^{1/p'}$$

$$\le C B_0 |A_1|^{1/p_1} \left(\frac{\sqrt{B_1}}{\sqrt{B_0}} \right)^{1/p_1'} |A_2|^{1/p_2} \left(\frac{\sqrt{B_2}}{\sqrt{B_0}} \right)^{1/p_2'} |A_0|^{1/p_1' + 1/p_2' - 1}$$

$$+ M \, 2^{-1/p'} |A_1|^{1/p_1} |A_2|^{1/p_2} |A_0|^{1/p'}.$$

It follows that M has to be less than or equal to

$$C B_0 \left(\frac{\sqrt{B_1}}{\sqrt{B_0}} \right)^{1/p_1'} \left(\frac{\sqrt{B_2}}{\sqrt{B_0}} \right)^{1/p_2'} + M \, 2^{-1/p'}$$

and consequently

$$M \leq \frac{C}{1 - 2^{-1/p'}} B_0^{1/(2p)} B_1^{1/(2p_1')} B_2^{1/(2p_2')}.$$

Case 2: Suppose that

$$\frac{|A_1|}{\sqrt{B_1}} \geq \max\left(\frac{|A_0|}{\sqrt{B_0}}, \frac{|A_2|}{\sqrt{B_2}}\right).$$

Here we use that T^{*1} maps $L^1 \times L^1$ to weak $L^{1/2}$ when restricted to characteristic functions. Then there exists a subset A_1' of A_1 of measure $|A_1'| \geq \frac{1}{2}|A_1|$ such that

$$\left|\int_{A_1'} T^{*1}(\chi_{A_0}, \chi_{A_2}) \, dx\right| \leq C \, B_1 \, |A_0| \, |A_2| \, |A_1|^{-1}$$

for some constant C. Equivalently, we write this statement as

$$\left|\int_{A_0} T(\chi_{A_1'}, \chi_{A_2}) \, dx\right| \leq C \, B_1 \, |A_0| \, |A_2| \, |A_1|^{-1}$$

by the definition of the first dual operator T^{*1}. Therefore we obtain

$$\left|\int_{A_0} T(\chi_{A_1}, \chi_{A_2}) \, dx\right| \leq \left|\int_{A_0} T(\chi_{A_1'}, \chi_{A_2}) \, dx\right| + \left|\int_{A_0} T(\chi_{A_1 \setminus A_1'}, \chi_{A_2}) \, dx\right|$$

$$\leq C \, B_1 \, |A_0| \, |A_2| \, |A_1|^{-1} + M|A_0|^{1/p'} |A_2|^{1/p_2} \left(\frac{1}{2}|A_1|\right)^{1/p_1}$$

$$\leq C \, B_1 \, |A_1|^{-1+1/p+1/p_2'} \left(\frac{\sqrt{B_0}}{\sqrt{B_1}}\right)^{1/p} |A_2|^{1/p_2} \left(\frac{\sqrt{B_2}}{\sqrt{B_1}}\right)^{1/p_2'} |A_0|^{1/p'}$$

$$+ M \, 2^{-1/p_1} |A_1|^{1/p_1} |A_2|^{1/p_2} |A_0|^{1/p'}.$$

By the definition of M, it follows that

$$M \leq \frac{C}{1 - 2^{-1/p_1}} B_0^{1/(2p)} B_1^{1/(2p_1')} B_2^{1/(2p_2')}.$$

Case 3: Suppose that

$$\frac{|A_2|}{\sqrt{B_2}} \geq \max\left(\frac{|A_0|}{\sqrt{B_0}}, \frac{|A_1|}{\sqrt{B_1}}\right);$$

here we proceed as in Case 2 with the roles of A_1 and A_2 interchanged.

Then the statement of the theorem follows with

$$C_{p_1,p_2} = C \max\left(\frac{1}{1 - 2^{-1/p_1}}, \frac{1}{1 - 2^{-1/p_2}}, \frac{1}{1 - 2^{-1/p'}}\right). \qquad \square$$

Assumption (2.18) is not as restrictive as it looks. To apply this theorem for bilinear Calderón–Zygmund operators, one needs to consider the family of operators whose kernels are truncated near the origin, i.e.,

$$K_\delta(x, y_1, y_2) = K(x, y_1, y_2)\, \widetilde{\chi}\big((|x - y_1| + |x - y_2|)/\delta\big),$$

where $\widetilde{\chi}$ is a smooth function that is equal to 1 on $[2, \infty)$ and vanishes on $[0, 1]$. The kernels K_δ are essentially in the same Calderón–Zygmund kernel class as K, that is, if K lies in 2-CZK(A, ε), then K_ε lie in 2-CZK(A', ε), where A' is a multiple of A. Using Hölder's inequality with exponents p_1, p_2, p', it is easy to see that for the operators T_δ with kernels K_δ, the assumption (2.18) holds with constants depending on δ.

Theorem 2.7 provides an interpolation machinery needed to pass from bounds at one point to bounds at every point for bilinear Calderón–Zygmund operators. (An alternative interpolation technique was described in [22].) We have:

Theorem 2.8. *Suppose that a bilinear operator T with kernel in 2-CZK(A, δ) and all of its truncations T_δ map $L^{r_1} \times L^{r_2}$ to L^r for a single triple of indices r_1, r_2, r satisfying $1/r_1 + 1/r_2 = 1/r$ and $1 < r_1, r_2, r < \infty$ uniformly in δ. Then T is bounded from $L^{p_1} \times L^{p_2}$ to L^p for all indices p_1, p_2, p satisfying $1/p_1 + 1/p_2 = 1/p$ and $1 < p_1, p_2 \leq \infty$, $1/2 < p < \infty$.*

Proof. Since T_δ maps $L^{r_1} \times L^{r_2} \to L^r$ and $r > 1$, duality gives that T_δ^{*1} maps $L^{r'} \times L^{r_2} \to L^{r_1'}$ and T_δ^{*2} maps $L^{r_1} \times L^{r'} \to L^{r_2'}$ (uniformly in δ). It follows from Theorems 2.6 and 2.7 that T_δ are bounded from $L^{p_1} \times L^{p_2} \to L^p$ for all indices p_1, p_2, p satisfying $1/p_1 + 1/p_2 = 1/p$ and $1 < p_1, p_2 < \infty$, $1/2 < p < \infty$. Passing to the limit, using Fatou's lemma, the same conclusion may be obtained for the nontruncated operator T. The case $p_1 = \infty$ or $p_2 = \infty$ follows by duality from the case $p = 1$.   □

The only drawback of this approach is that it is based on the assumption that if T is bounded from $L^{r_1} \times L^{r_2} \to L^r$, then so are all its truncations T_δ (uniformly in $\delta > 0$). This is hardly a problem in concrete applications since the kernels of T and T_δ satisfy equivalent estimates (uniformly in $\delta > 0$) and the method used in the proof of boundedness of the former almost always applies for the latter.

Proposition 2.9. *Let T be a bilinear operator associated with a kernel of class 2-CZK(A, ε) that maps $L^{q_1} \times L^{q_2}$ to L^q for some choice of indices q_1, q_2, q that satisfy $1/q_1 + 1/q_2 = 1/q$. Then T has an extension that maps*

$$L^\infty \times L^\infty \longrightarrow BMO$$

with bound a constant multiple of $A + B$. By duality, T also maps

$$L^\infty \times H^1 \longrightarrow L^1, \quad \text{and also} H^1 \times L^\infty \longrightarrow L^1,$$

where H^1 is the Hardy 1-space.

The proof is based on a straightforward adaptation of the Peetre, Spanne, and Stein result [35], [39], [40] on boundedness of a linear Calderón–Zygmund operator from L^∞ to BMO, and is omitted.

2.4 The bilinear $T1$ theorem

In this section we quickly discuss the bilinear version of the $T1$ theorem. The scope of this theorem is to provide a sufficient condition for boundedness of bilinear Calderón–Zygmund operators at some point, i.e., $L^{q_1} \times L^{q_2} \to L^{q,\infty}$ for some choice of points q_1, q_2, q. Once this is known, then the bilinear version of the Calderón–Zygmund theorem (Theorem 2.4) combined with interpolation yields boundedness of these operators in the entire range of indices where boundedness is possible.

The linear $T1$ theorem was obtained by David and Journé [10]. Its original formulation involves three conditions equivalent to L^2 boundedness. These conditions are that $T1 \in BMO$, $T^*1 \in BMO$, and that a certain weak boundedness property holds. This version of the David–Journé $T1$ theorem was extended to the multilinear setting by Christ and Journé [5]. Another version of the $T1$ theorem using exponentials appears in [10] and is better suited for our purposes. This version is as follows: A linear operator T with kernel in the Calderón–Zygmund class $\text{CZK}(A, \varepsilon)$ maps $L^2(\mathbf{R}^n) \to L^2(\mathbf{R}^n)$ if and only if

$$\sup_{\xi \in \mathbf{R}^n} \left(\|T(e^{2\pi i \xi \cdot (\cdot)})\|_{BMO} + \|T^*(e^{2\pi i \xi \cdot (\cdot)})\|_{BMO} \right) < \infty.$$

In this section we will state and prove a multilinear version of the $T1$ theorem using the characterization stated above.

Theorem 2.10. *Fix $1 < q_1, q_2, q < \infty$ with*

$$\frac{1}{q_1} + \frac{1}{q_2} = \frac{1}{q}. \tag{2.19}$$

Let T be a continuous bilinear operator from $\mathcal{S}(\mathbf{R}^n) \times \mathcal{S}(\mathbf{R}^n) \to \mathcal{S}'(\mathbf{R}^n)$ with kernel K in $2\text{-}\text{CZK}(A, \varepsilon)$. Then T has a bounded extension from $L^{q_1} \times L^{q_2}$ to L^q if and only if

$$\sup_{\xi_1 \in \mathbf{R}^n} \sup_{\xi_2 \in \mathbf{R}^n} \|T(e^{2\pi i \xi_1 \cdot (\cdot)}, e^{2\pi i \xi_2 \cdot (\cdot)})\|_{BMO} \le B \tag{2.20}$$

and also

$$\sup_{\xi_1 \in \mathbf{R}^n} \sup_{\xi_2 \in \mathbf{R}^n} \|T^{*j}(e^{2\pi i \xi_1 \cdot (\cdot)}, e^{2\pi i \xi_2 \cdot (\cdot)})\|_{BMO} \le B \tag{2.21}$$

for all $j = 1, 2$. Moreover, if (2.20) and (2.21) hold then we have that

$$\|T\|_{L^{q_1} \times L^{q_2} \to L^q} \le c_{n,q_1,q_2}(A + B),$$

for some constant c_{n,m,q_j} depending only on the parameters indicated.

Proof. Obviously the necessity of conditions (2.20) and (2.21) follows from Proposition 2.9. The thrust of this theorem is provided by the sufficiency of these conditions, i.e., the fact that, if (2.20) and (2.21) hold, then T extends to a bounded operator from $L^{q_1} \times L^{q_2}$ to L^q. Although we are not going to be precise in the proof of this result, we make some comments. The outline of the proof is based on another formulation of the $T1$ theorem given by Stein [42]. Let us consider the set of all C^∞ functions supported in the unit ball of $\mathbf{R}^n$ satisfying

$$\|\partial^\alpha \phi\|_{L^\infty} \leq 1$$

for all multiindices $|\alpha| \leq [n/2] + 1$. Such functions are called *normalized bumps*. For a normalized bump ϕ, $x_0 \in \mathbf{R}^n$, and $R > 0$, define the function

$$\phi^{R,x_0}(x) = \phi\left(\frac{x - x_0}{R}\right).$$

The formulation in [42, Theorem 3, page 294] says that a necessary and sufficient condition for an operator T with kernel of class $CZK(A, \varepsilon)$ to be L^2-bounded is that for some constant $B > 0$ we have

$$\|T(\phi^{R,x_0})\|_{L^2} + \|T^*(\phi^{R,x_0})\|_{L^2} \leq B\, R^{n/2}$$

for all normalized bumps ϕ, all $R > 0$ and all $x_0 \in \mathbf{R}^n$. Moreover, the norm of the operator T on L^2 (and therefore on L^p) is bounded by a constant multiple of $A + B$. Adopting this terminology in the bilinear setting, we say that a bilinear operator T is *BMO-restrictedly bounded with bound C* if

$$\|T(\phi_1^{R_1,x_1}, \phi_2^{R_2,x_2})\|_{BMO} \leq C < \infty$$

and

$$\|T^{*j}(\phi_1^{R_1,x_1}, \phi_2^{R_2,x_2})\|_{BMO} \leq C < \infty$$

for all $j = 1, 2$, all normalized bumps ϕ_j, all $R_j > 0$, and all $x_j \in \mathbf{R}^n$.

The main observation is that if (2.20) and (2.21) are satisfied, then T is BMO-restrictedly bounded with bound a multiple of $B > 0$. This observation can be obtained from the corresponding result for linear operators. Consider the linear operator

$$T_{\phi_2^{R_2,x_2}}(f_1) = T(f_1, \phi_2^{R_2,x_2})$$

obtained from T by freezing an arbitrary normalized bump in the second entry. It is easy to see that $T_{\phi_2^{R_2,x_2}}$ satisfies the linear BMO-restrictedly boundedness condition with bound B. It follows from this that $T_{\phi_m^{R_m,x_m}}$ maps the space of bounded functions with compact support to BMO with norm at most a multiple of $A + B$, i.e.,

$$\|T(g, \phi_2^{R_2,x_2})\|_{BMO} \leq c(A + B)\|g\|_{L^\infty} \tag{2.22}$$

holds for bounded functions g with compact support.

Now consider the operators T_{g_1} defined by

$$T_{g_1}(f_2) = T(g_1, f_2),$$

for a compactly supported and bounded function g_1. The estimate (2.22) is saying that T_{g_1} satisfies the linear BMO-restrictedly boundedness condition with constant a multiple of $(A + B)\|g_1\|_{L^\infty}$. The corresponding linear result implies that

$$T: L^q \times L_c^\infty \longrightarrow L^q,$$

with norm controlled by a multiple of $A + B$. (Here L_c^∞ is the space of bounded functions with compact support.) Furthermore, this result can be used as a starting point for the boundedness of T from $L^1 \times L^1$ to $L^{1/2,\infty}$, which in turn implies boundedness of T in the range $L^{q_1} \times L^{q_2} \to L^q$ for all $1 < q_1, q_2 \leq \infty$, $1/2 < q < \infty$, in view of Theorem 2.8. □

As an application, we obtain a bilinear extension of a result of Bourdaud [2].

Example 2.11. Consider the class of bilinear pseudodifferential operators

$$T(f_1, f_2)(x) = \int_{\mathbf{R}^n} \int_{\mathbf{R}^n} \sigma(x, \xi_1, \xi_2) \, \widehat{f_1}(\xi_1) \, \widehat{f_2}(\xi_2) \, e^{2\pi i x \cdot (\xi_1 + \xi_2)} \, d\xi_1 \, d\xi_2$$

with symbols σ satisfying

$$|\partial_x^\alpha \partial_{\xi_1}^{\beta_1} \partial_{\xi_2}^{\beta_2} \sigma(x, \xi_1, \xi_2)| \leq C_{\alpha, \beta}(1 + |\xi_1| + |\xi_2|)^{|\alpha| - (|\beta_1| + |\beta_2|)}, \qquad (2.23)$$

for all n-tuples α, β_1, β_2 of nonnegative integers. It is easy to see that such operators have kernels in 2-CZK. For these operators we have that

$$T\big(e^{2\pi i \eta_1 \cdot (\,\cdot\,)}, e^{2\pi i \eta_2 \cdot (\,\cdot\,)}\big) = \sigma(x, \eta_1, \eta_2) \, e^{2\pi i x \cdot (\eta_1 + \eta_2)},$$

which is uniformly bounded in $\eta_j \in \mathbf{R}^n$. Theorem 2.10 implies that a necessary and sufficient condition for T to map a product of L^p spaces into another Lebesgue space with the usual relation on the indices is that $T^{*j}(e^{2\pi i \eta_1 \cdot (\,\cdot\,)}, e^{2\pi i \eta_2 \cdot (\,\cdot\,)})$ are in BMO uniformly in $\eta_k \in \mathbf{R}^n$. In particular, this is the case if all the transposes of T have symbols that also satisfy (2.23).

We now look for sufficient conditions on a singular kernel K_0 so that the corresponding translation invariant operator

$$T(f_1, f_2)(x) = \int_{\mathbf{R}^n} \int_{\mathbf{R}^n} K_0(x - y_1, x - y_2) \, f_1(y_1) \, f_2(y_2) \, dy_1 \, dy_2 \qquad (2.24)$$

maps $L^{p_1} \times L^{p_2}$ to L^p when the indices satisfy $1/p_1 + 1/p_2 = 1/p$. We have the following:

Theorem 2.12. *Let $K_0(u_1, u_2)$ be a locally integrable function on $(\mathbf{R}^n)^2 \setminus \{(0,0)\}$ which satisfies the size estimate*

$$|K_0(u_1, u_2)| \leq A|(u_1, u_2)|^{-2n}, \qquad (2.25)$$

the cancellation condition

$$\left| \iint_{R_1 < |(u_1, u_2)| < R_2} K_0(u_1, u_2) \, du_1 \, du_2 \right| \leq A < \infty \qquad (2.26)$$

for all $0 < R_1 < R_2 < \infty$, and the smoothness condition

$$|K_0(u_1, u_2) - K_0(u_1, u'_2)| \le A \frac{|u_2 - u'_2|^\varepsilon}{|(u_1, u_2)|^{2n+\varepsilon}} \tag{2.27}$$

whenever $|u_2 - u'_2| < \frac{1}{2}|u_2|$. Then the multilinear operator T given by (2.24) maps $L^{p_1} \times L^{p_2}$ into L^p when $1 < p_j < \infty$ and (2.29) is satisfied. In particular, this is the case if K_0 has the form

$$K_0(u_1, u_2) = \frac{\Omega\left(\dfrac{(u_1, u_2)}{|(u_1, u_2)|}\right)}{|(u_1, u_2)|^{2n}}$$

and Ω is an integrable function with mean value zero on the sphere $\mathbf{S}^{2n-1}$ which is Lipschitz of order $\varepsilon > 0$.

Proof. This theorem is a consequence of Theorem 2.10. As in the previous application of this theorem (with some formal computations that are easily justified), we have

$$T\big(e^{2\pi i \eta_1 \cdot (\,\cdot\,)}, e^{2\pi i \eta_2 \cdot (\,\cdot\,)}\big)(x) = e^{2\pi i x \cdot (\eta_1 + \eta_2)}\, \widehat{K_0}(\eta_1, \eta_2),$$

which is a bounded function, hence in BMO. The fact that $\widehat{K_0}$ is bounded is a standard fact; see for instance [1]. And certainly the same result is valid for all smooth truncations of K_0, a condition needed in the application of Theorem 2.8. The calculations with the transposes are similar; for example,

$$T^{*1}\big(e^{2\pi i \eta_1 \cdot (\,\cdot\,)}, e^{2\pi i \eta_2 \cdot (\,\cdot\,)}\big)(x) = e^{2\pi i x \cdot (\eta_1 + \eta_2)}\, \widehat{K_0}(-\eta_1 - \eta_2, \eta_2),$$

which is also in BMO. $\qquad\qquad\qquad\qquad\qquad\qquad\qquad\qquad\qquad\qquad\qquad\square$

Another application of this result concerns the bilinear extension of the classical Hörmander–Mihlin multiplier theorem discussed in the Introduction. The multilinear analogue of the Hörmander–Mihlin multiplier theorem was obtained by Coifman and Meyer [6], [7] on L^p for $p \ge 1$ and extended to indices $p < 1$ in [22] and [27].

Theorem 2.13. *Suppose that $a(\xi_1, \xi_2)$ is a C^∞ function on $(\mathbf{R}^n)^2 \setminus \{(0,0)\}$ which satisfies*

$$|\partial_{\xi_1}^{\beta_1} \partial_{\xi_2}^{\beta_2} a(\xi_1, \xi_2)| \le C_{\beta_1, \beta_2}(|\xi_1| + |\xi_2|)^{-(|\beta_1|+|\beta_2|)} \tag{2.28}$$

for all multiindices β_1, β_2. Then the corresponding bilinear operator T with symbol a is a bounded operator from $L^{p_1} \times L^{p_2}$ into L^p when $1 < p_j < \infty$ and

$$\frac{1}{p_1} + \frac{1}{p_2} = \frac{1}{p}, \tag{2.29}$$

and it also maps $L^1 \times L^1$ to $L^{1/2, \infty}$.

Proof. Indeed, conditions (2.28) easily imply that the inverse Fourier transform of a satisfies

$$|\partial_{\xi_1}^{\beta_1}\partial_{\xi_2}^{\beta_2}a^\vee(x_1,x_2)| \leq C_{\beta_1,\beta_2}(|x_1|+|x_2|)^{-(2n+|\beta_1|+|\beta_2|)} \qquad (2.30)$$

for all multiindices β_1, β_2. It follows that the kernel

$$K(x,y_1,y_2) = a^\vee(x-y_1,x-y_2)$$

of the operator T satisfies the required size and smoothness conditions (2.6), (2.7), and (2.8). The $L^{q_1} \times L^{q_2} \to L^q$ boundedness of T for a fixed point $(1/q_1,1/q_2,1/q)$ follows from the bilinear $T1$ theorem (Theorem 2.10) once we have verified the required BMO conditions. As in previous examples, we have that

$$T\big(e^{2\pi i\eta_1\cdot(\,\cdot\,)}, e^{2\pi i\eta_2\cdot(\,\cdot\,)}\big)(x) = a(\eta_1,\eta_2)\, e^{2\pi ix\cdot(\eta_1+\eta_2)},$$

which is in L^∞ and thus in BMO uniformly in η_1,η_2. The same calculation is valid for the two transposes of T, since their corresponding multipliers also satisfy (2.28). The weak type results follow from Theorem 2.4. $\qquad\square$

2.5 Orthogonality properties for bilinear multiplier operators

In view of Plancherel's theorem, the space L^2 turns into a heaven of orthogonality since Fourier multiplier operators turn into multiplication operators. There is no analogous phenomenon in the bilinear setting, but there is a range of indices for which bilinear multiplier operators exhibit properties analogous to those of linear multiplier operators on L^2. The situation where $2 \leq p_1,p_2,p' < \infty$, $1/p_1 + 1/p_2 = 1/p$ is called the *local L^2 case*. There is a noteworthy orthogonality lemma for bilinear operators in the local L^2 case. The one-dimensional version of the result below appeared in [18].

Theorem 2.14. *Suppose that T_j, $j \in \mathbf{Z}$, are bilinear operators whose symbols $m_j(\xi_1,\xi_2)$ are supported in sets $A_j \times B_j$, where $\{A_j\}_j$ is a family of pairwise disjoint rectangles on $\mathbf{R}^n$ and $\{B_j\}_j$ is also a family of pairwise disjoint rectangles on $\mathbf{R}^n$. Assume, furthermore, that $\{A_j + B_j\}_j$ is a family of pairwise disjoint rectangles on $\mathbf{R}^n$ and that for some indices p_1,p_2,p in the local L^2 case one has that the bilinear operators*

$$T_{m_j}(f,g)(x) = \int_{\mathbf{R}^n}\int_{\mathbf{R}^n} m_j(\xi,\eta)\,\widehat{f}(\xi)\,\widehat{g}(\eta)\, e^{2\pi ix\cdot(\xi+\eta)}\,d\xi\,d\eta$$

with symbols m_j are uniformly bounded, that is,

$$\sup_{j\in\mathbf{Z}} \big\|T_{m_j}\big\|_{L^{p_1}\times L^{p_2}\to L^p} \leq C < \infty.$$

whenever p_1, p_2, p are indices in the local L^2 case, i.e., $2 \le p_1, p_2, p' < \infty$. Then there is a finite constant $C_{p_1, p_2, n}$ such that

$$\left\| \sum_{j \in \mathbf{Z}} T_{m_j} \right\|_{L^{p_1} \times L^{p_2} \to L^p} \le C_{p_1, p_2, n} \, C, \tag{2.31}$$

that is, the sum of the T_{m_j} is also a bounded operator in the local L^2 case.

Proof. The key element of the proof of this result is the nth-dimensional version of Rubio de Francia's [36] Littlewood–Paley inequality for arbitrary disjoint intervals. This says that, if R_j is a family of disjoint rectangles in $\mathbf{R}^n$ and if $\Delta_j f = (\hat{f} \chi_{R_j})^\vee$, then there is a constant $c_{p,n}$ such that for $2 \le p < \infty$ and all functions $f \in L^p(\mathbf{R}^n)$ one has

$$\left\| \left(\sum_{j \in \mathbf{Z}} |\Delta_j f|^2 \right)^{1/2} \right\|_{L^p(\mathbf{R}^n)} \le c_{p,n} \, \|f\|_{L^p(\mathbf{R}^n)}. \tag{2.32}$$

This result was proved by Journé [26], but easier proofs of it were later provided by Soria [38] in dimension 2 and Sato [37] in higher dimensions.

We will prove (2.31) by duality. We denote by $p' = p/(p-1)$ the dual index of p. We introduce Littlewood–Paley operators

$$\Delta_j^1 f = (\hat{f} \chi_{A_j})^\vee, \qquad \Delta_j^2 g = (\hat{g} \chi_{A_j})^\vee. \qquad \Delta_j^3 h = (\hat{h} \chi_{A_j + B_j})^\vee.$$

An easy calculation shows that

$$\widehat{T_{m_j}(f,g)}(\xi) = \int_{\mathbf{R}^n} m_j(\eta, \xi - \eta) \, \hat{f}(\eta) \, \hat{g}(\xi - \eta) \, d\eta; \tag{2.33}$$

thus the Fourier transform of $T_{m_j}(f,g)$ is supported in the set $A_j + B_j$. Then we have

$$\langle T_{m_j}(f,g), h \rangle = \langle T_{m_j}(\Delta_j^1 f, \Delta_j^2 g), \Delta_j^3 h \rangle$$

in view of the hypotheses on the m_j's. Consequently,

$$\left| \left\langle \sum_j T_{m_j}(f,g), h \right\rangle \right| = \left| \sum_j \langle T_{m_j}(\Delta_j^1 f, \Delta_j^2 g), \Delta_j^3 h \rangle \right|$$

$$\le \int_{\mathbf{R}^n} \left(\sum_j |T_{m_j}(\Delta_j^1 f, \Delta_j^2 g)|^2 \right)^{1/2} \left(\sum_j |\Delta_j^3 h|^2 \right)^{1/2} dx$$

$$\le \left\| \left(\sum_j |T_{m_j}(\Delta_j^1 f, \Delta_j^2 g)|^2 \right)^{1/2} \right\|_{L^p} \left\| \left(\sum_j |\Delta_j^3 h|^2 \right)^{1/2} \right\|_{L^{p'}}$$

$$\le \left\| \left(\sum_j |T_{m_j}(\Delta_j^1 f, \Delta_j^2 g)|^2 \right)^{1/2} \right\|_{L^p} c_{p',n} \|h\|_{L^{p'}},$$

where the last inequality follows from (2.32) since $p' \geq 2$. It suffices to estimate the square function above. Since $p/2 \leq 1$, we have the first inequality below:

$$\left\| \left(\sum_j |T_{m_j}(\Delta_j^1 f, \Delta_j^2 g)|^2 \right)^{1/2} \right\|_{L^p}^p \leq \int_{\mathbf{R}^n} \sum_j |T_{m_j}(\Delta_j^1 f, \Delta_j^2 g)|^p \, dx$$

$$= \sum_j \|T_{m_j}(\Delta_j^1 f, \Delta_j^2 g)\|_{L^p}^p$$

$$\leq C^p \sum_j \|\Delta_j^1 f\|_{L^{p_1}}^p \|\Delta_j^2 g\|_{L^{p_2}}^p$$

$$\leq C^p \Big(\sum_j \|\Delta_j^1 f\|_{L^{p_1}}^{p_1} \Big)^{p/p_1} \Big(\sum_j \|\Delta_j^2 g\|_{L^{p_2}}^{p_2} \Big)^{p/p_2}$$

$$\leq C^p \left\| \Big(\sum_j |\Delta_j^1 f|^2 \Big)^{1/2} \right\|_{L^{p_1}}^p \left\| \Big(\sum_j |\Delta_j^2 g|^2 \Big)^{1/2} \right\|_{L^{p_2}}^p$$

$$\leq (C c_{p_1,n} c_{p_2,n})^p \|f\|_{L^{p_1}}^p \|g\|_{L^{p_2}}^p,$$

where we used successively the uniform boundedness of the T_{m_j}'s, Hölder's inequality, and (2.32). The required conclusion follows with $C_{p_1,p_2,n} = c_{p_1,n} c_{p_2,n} c_{p',n}$. (Recall that $1/p = 1/p_1 + 1/p_2$; thus the dependence of the constant on p is expressed via that of p_1 and p_2.) $\qquad \square$

This result was used in [18] in proving that the characteristic function of the unit disc is a bounded bilinear multiplier in the local L^2 case. It also has other applications; below we use this result to provide another proof of the Coifman–Meyer multiplier theorem (Theorem 2.13) for bilinear operators. In view of the bilinear Calderón–Zygmund theorem, it suffices to prove boundedness at a single point; we pick this point to be $L^{p_1} \times L^{p_2} \to L^2$, where $1/p_1 + 1/p_2 = 1/2$ and we only prove it for a piece of the operator, as the remaining pieces are handled by duality.

We have the following:

Theorem 2.15. *Suppose that $m(\xi, \eta)$ is a function on $\mathbf{R}^n \times \mathbf{R}^n$ that satisfies*

$$|\partial_\xi^\alpha \partial_\eta^\beta m(\xi, \eta)| \leq C_{\alpha,\beta}(|\xi| + |\eta|)^{-|\alpha|-|\beta|} \tag{2.34}$$

for multiindices $|\alpha|, |\beta| \leq 2n + 1$. Then the multiplier operator

$$T_m(f_1, f_2)(x) = \int_{\mathbf{R}^n} \int_{\mathbf{R}^n} \widehat{f_1}(\xi_1) \widehat{f_2}(\xi_2) \, m(\xi_1, \xi_2) \, e^{2\pi i x \cdot (\xi_1 + \xi_2)} \, d\xi_1 \, d\xi_2$$

maps $L^{p_1} \times L^{p_2}$ to L^p, where $1/p_1 + 1/p_2 = 1/p$ and $1 < p_1, p_2, p < \infty$.

Proof. We begin by making the observation that T_m^{*1} is a bilinear multiplier operator with symbol

$$m^{*1}(\xi, \eta) = m(-\xi - \eta, \eta)$$

and T_m^{*2} is a bilinear multiplier operator with symbol

$$m^{*2}(\xi, \eta) = m(\xi, -\xi - \eta).$$

Both of these symbols satisfy condition (2.34) for all multiindices $|\alpha|, |\beta| \leq n + 2$ (for some other constants). Moreover, if $m(\xi, \eta)$ is supported near the diagonal $|\xi| \approx |\eta|$, then $m^{*1}(\xi, \eta)$ is supported near the axis $|\xi| \lesssim |\eta|$ and $m^{*2}(\xi, \eta)$ is supported near the axis $|\eta| \lesssim |\xi|$. We introduce a smooth partition of unity on the sphere $\mathbf{S}^{2n-1}$ such that $1 = \phi_0 + \phi_1 + \phi_2 + \phi_3$, where $\phi_0(\xi, \eta)$ is supported in a neighborhood of the diagonal $|\xi| \approx |\eta| \approx |\xi + \eta|$, ϕ_1 is supported in a set of the form $|\xi| \lesssim |\eta|$, ϕ_2 is supported in a set of the form $|\eta| \lesssim |\xi|$, and ϕ_3 is supported in a set of the form $|\xi| \approx |\eta| \approx |\xi - \eta|$. We extend the functions ϕ_j so that they be homogeneous on $\mathbf{R}^{2n}$. We split the multiplier m as

$$m = m\,\phi_0 + m\,\phi_1 + m\,\phi_2 + m\,\phi_3 =: m_0 + m_1 + m_2 + m_3$$

and we say that m_0 is supported in the "good" direction, while m_1, m_2, m_3 are supported in the remaining three "bad" directions. We observe that the three bad directions are "preserved" by duality. By this we mean that $T_{m_3}^{*1}$ is an operator whose symbol has the same properties as m_1, and $T_{m_3}^{*2}$ is an operator whose symbol has the same properties as m_2.

We introduce a function ψ supported in the annulus $1/4 < |(\xi, \eta)| < 1 - 10^{-1}$ such that

$$\sum_{j \in \mathbf{Z}} \psi(2^{-j}(\xi, \eta)) = 1$$

for all $(\xi, \eta) \neq 0$. Then we write $m_r^j(\xi, \eta) = m_r(\xi, \eta)\psi(2^{-j}(\xi, \eta))$ and consider the operators $T_{m_r^j}$ with symbols m_r^j. A simple observation is that, for all $r \in \{0, 1, 2, 3\}$ and all $j \in \mathbf{Z}$, the symbols m_r, m_r^j, and m_r^0 satisfy (2.34) with constants $C'_{\alpha,\beta}$ independent of j and explicitly related to the constants $C_{\alpha,\beta}$ appearing in (2.34).

We first show that for all $r \in \{0, 1, 2, 3\}$ the operator $T_{m_r^j}$ maps $L^{p_1} \times L^{p_2}$ to L^2 when $1/p_1 + 1/p_2 = 1/2$ uniformly in j. In fact, it suffices to take $j = 0$, noting that a dilation argument reduces the general case to that of $j = 0$, since m_r^j satisfy (2.34) uniformly in j. Let us fix two Schwartz functions f_1 and f_2 on $\mathbf{R}^n$. We have

$$\left\| T_{m_r^0}(f_1, f_2) \right\|_{L^2}^2 = \int_{\mathbf{R}^n} \left| \int_{\mathbf{R}^n} \widehat{f_1}(\xi_1)\,\widehat{f_2}(\xi - \xi_1)\,m_r^0(\xi_1, \xi - \xi_1)\,d\xi_1 \right|^2 d\xi \qquad (2.35)$$

in view of (2.33). Since m_r^0 is supported in $[-1, 1]^{2n}$, we expand it in Fourier series as follows:

$$m_r^0(\xi_1, \xi_2) = \sum_{k,l \in \mathbf{Z}^n} c_{k,l}\, e^{2\pi i k \cdot \xi_1}\, e^{2\pi i l \cdot \xi_2}\, \zeta(\xi_1)\, \zeta(\xi_2), \qquad (2.36)$$

where ζ is a smooth function supported in the annulus $1/8 < |\xi| < 1$ and is equal to 1 on the support of ψ. Here $c_{k,l}$ are the Fourier coefficients of the expansion given by

$$c_{k,l} = \int_A \int_A m_0^0(t_1, t_2)\, e^{-2\pi i k \cdot t_1}\, e^{-2\pi i l \cdot t_2}\, dt_1\, dt_2,$$

where A is the n-dimensional annulus $1/4 < |\xi| < 1$. To obtain estimates for $c_{k,l}$, we integrate by parts with respect to the differential operators $(I - \Delta_{t_1})^L (I - \Delta_{t_2})^L$, where Δ_{t_1} is the Laplacian with respect to the variable t_1 and $L = [n/2] + 1$. One obtains

$$c_{k,l} = \frac{\displaystyle\int_A \int_A \left\{ (I - \Delta_{t_1})^L (I - \Delta_{t_2})^L m_0^0(t_1, t_2) \right\} e^{-2\pi i k \cdot t_1}\, e^{-2\pi i l \cdot t_2}\, dt_1\, dt_2}{(1 + 4\pi^2 |k|^2)^L\, (1 + 4\pi^2 |k|^2)^L}.$$

Condition (2.34) implies that the integrand above is uniformly bounded in k, l. Consequently we have the estimate

$$|c_{k,l}| \leq \frac{C_n \sum_{|\alpha|, |\beta| \leq L} C_{\alpha, \beta}}{(1 + 4\pi^2 |k|^2)^L\, (1 + 4\pi^2 |k|^2)^L}$$

and, since $2L > n$, we deduce that

$$\sum_{k,l \in \mathbf{Z}^n} c_{k,l} < C_n' \sum_{|\alpha|, |\beta| \leq L} C_{\alpha, \beta} < \infty.$$

Using this fact and (2.35) and (2.36), we obtain

$$\left\| T_{m_r^0}(f_1, f_2) \right\|_{L^2}^2 \leq \int_{\mathbf{R}^n} \sum_{k,l \in \mathbf{Z}^n} c_{k,l} \left| \int_{\mathbf{R}^n} \widehat{f_1}(\xi_1)\, \widehat{f_2}(\xi - \xi_1)\, \zeta_k(\xi_1)\, \zeta_l(\xi - \xi_1)\, d\xi_1 \right|^2 d\xi,$$

where $\zeta_k(t) = \zeta(t) e^{2\pi i k \cdot t}$ and $\zeta_l(t) = \zeta(t) e^{2\pi i l \cdot t}$. Let S_{ζ_k} be the linear multiplier operator $S_{\zeta_k} f = \left(\widehat{f}\, \zeta_k \right)^\vee$. Clearly the last expression above is equal to

$$\sum_{k,l \in \mathbf{Z}^n} c_{k,l} \int_{\mathbf{R}^n} \left| \widehat{S_{\zeta_k} f_1} * \widehat{S_{\zeta_l} f_2} \right|^2 d\xi = \sum_{k,l \in \mathbf{Z}^n} c_{k,l} \int_{\mathbf{R}^n} \left| S_{\zeta_k} f_1(x)\, S_{\zeta_l} f_2(x) \right|^2 dx.$$

By Hölder's inequality, this expression is certainly bounded by

$$\sum_{k,l \in \mathbf{Z}^n} c_{k,l} \left\| S_{\zeta_k} f_1 \right\|_{L^{p_1}}^2 \left\| S_{\zeta_l} f_2 \right\|_{L^{p_2}}^2$$

and the latter is clearly at most a constant multiple of $\| f_1 \|_{L^{p_1}}^2 \| f_2 \|_{L^{p_2}}^2$.

Now we turn to the sum over $j \in \mathbf{Z}$ for each family $\{ m_r^j \}_j$. As we pointed out earlier, we have that, for each r, $T_{m_r^j}$ maps $L^{p_1} \times L^{p_2} \to L^2$, where $1/p_1 + 1/p_2 = 1/2$ uniformly in j. As a consequence of (2.34), it follows that the kernel $K(x)$ of each $T_{m_r^j}$ and each T_{m_r} satisfies

$$|K(x_1, x_2)| + |(x_1, x_2)|\, |\nabla K(x_1, x_2)| \leq C |(x_1, x_2)|^{-2n}$$

uniformly in j, thus it is of class 2-CZK$(A, 1)$. By Theorem 2.4, it follows that $T_{m_r^j}$ maps $L^1 \times L^1 \to L^{1/2,\infty}$. It follows from Theorem 2.6 that $T_{m_r^j}$ maps $L^{p_1} \times L^{p_2} \to L^p$, where $1/p_1 + 1/p_2 = 1/p$ and $2 \le p_1, p_2, p' \le \infty$. Thus each $T_{m_r^j}$ is bounded on the closure of the local L^2 triangle. We may apply Theorem 2.14 to obtain boundedness of the part of the operator corresponding to the good part of the symbol m_0, i.e.,

$$m_0(\xi, \eta) = \sum_{j \in \mathbf{Z}} m_0^j(\xi, \eta),$$

since the supports of the projections of the m_0^j's are pairwise disjoint (if they are split up in families of 10 elements indexed by numbers mod 10).

Unfortunately, we may not use orthogonality in the entire local L^2 triangle to obtain the same conclusion for the bad directions corresponding to the symbols m_1^j, m_2^j, and m_3^j. We note, however, that orthogonality can be used to obtain boundedness of T_{m_3} from that of each $T_{m_3^j}$ at the point $L^2 \times L^2 \to L^1$. This is in view of the observation that $T_{m_3^j}(f_1, f_2) = T_{m_3^j}(\Delta_j f_1, \Delta_j f_2)$ and of the next simple argument:

$$\left\| T_{m_3}(f_1, f_2) \right\|_{L^1} \le \sum_j \left\| T_{m_3^j}(\Delta_j f_1, \Delta_j f_2) \right\|_{L^1}$$

$$\le C \sum_j \left\| T_{m_3^j}(\Delta_j f_1, \Delta_j f_2) \right\|_{L^1}$$

$$\le C' \sum_j \left\| \Delta_j f_1 \right\|_{L^2} \left\| \Delta_j f_2 \right\|_{L^2}$$

$$\le C' \left(\sum_j \left\| \Delta_j f_1 \right\|_{L^2}^2 \right)^{\frac{1}{2}} \left(\sum_j \left\| \Delta_{kj} f_2 \right\|_{L^2}^2 \right)^{\frac{1}{2}}$$

$$\le C'' \left\| f_1 \right\|_{L^2} \left\| f_2 \right\|_{L^2}.$$

Here Δ_j is a Littlewood–Paley operator with Fourier transform localization near 2^j. Duality (or a similar orthogonality argument) implies boundedness of T_{m_1} from $L^\infty \times L^2$ to L^2 and analogously boundedness of T_{m_2} from $L^2 \times L^\infty$ to L^2. Bilinear complex interpolation between this estimate and the estimate $L^1 \times L^1 \to L^{1/2,\infty}$ gives that T_{m_2} maps $L^{3/2} \times L^3$ to the Lorentz space $L^{1,4/3}$ (thus to weak $L^{1,\infty}$). But we will need to work with a point near $(2/3, 1/3, 1)$ that still lies on the line that joins $(1/2, 0, 1/2)$ to $(1, 1, 2)$ and has third coordinate strictly smaller than 1, i.e., the target space is $L^{s,\infty}$ for some $s > 1$, which is a dual space. Since T_{m_2} is bounded at this point, it follows by duality that T_{m_3} is bounded from $L^{p,1} \times L^{q,1}$ to $L^{r,\infty}$ for some point $(1/p, 1/q, 1/r)$ near the point $(2/3, 0, 2/3)$ with $q < \infty$. Using Theorem 2.6 we obtain strong boundedness of T_{m_3} in the open triangle with vertices $(1, 1, 2)$, $(1/2, 1/2, 1)$, and $(2/3, 0, 2/3)$. In particular, it follows that T_{m_3} is bounded from $L^p \times L^q$ to L^r for some point $(1/p, 1/q, 1/r)$ near the point $(3/4, 1/4, 1)$. The preceding argument with $(3/4, 4, 1)$ in place of $(1/2, 1/2, 1)$ yields boundedness of T_{m_3} in yet a bigger region closer to the point $(1, 0, 1)$. Continuing this process indefinitely yields boundedness for T_{m_3} in the open quadrangle with corners $(1, 1, 2)$, $(1, 0, 1)$, $(2/3, 0, 2/3)$, and $(1/2, 1/2, 1)$.

A similar argument (involving the other adjoint) yields boundedness of T_{m_3} in the open quadrangle with corners $(1, 1, 2)$, $(0, 1, 1)$, $(0, 2/3, 2/3)$, and $(1/2, 1/2, 1)$. Further interpolation provides boundedness of T_{m_3} in the pentagon with vertices $(1, 1, 2)$, $(0, 2/3, 2/3)$, $(2/3, 0, 2/3)$, $(1, 1, 0)$, and $(0, 1, 1)$, that is, T_{m_3} is bounded from $L^p \times L^q$ to L^r whenever $r < 3/2$. (Recall that we always have $1/p + 1/q = 1/r$.)

Duality implies boundedness of T_{m_2} from $L^p \times L^q$ to L^r whenever $q > 3$ and that of T_{m_1} from $L^p \times L^q$ to L^r whenever $p > 3$. Further interpolation with the point $(1, 1, 2)$ yields boundedness of T_{m_1} and T_{m_2} in two open (convex) quadrilaterals Q_1 and Q_2 such that $Q_1 \cap Q_2$ has a nontrivial intersection with the region Q_3 described by the condition $r < 3/2$. The intersection $Q_1 \cap Q_2 \cap Q_3$ contains the point $(2/5, 2/5, 4/5)$ and thus T_{m_1}, T_{m_2}, and T_{m_3} are all bounded from $L^{5/2} \times L^{5/2} \to L^{5/4}$. We previously showed that T_{m_0} is bounded in the entire local L^2 triangle, thus T_m is also bounded from $L^{5/2} \times L^{5/2} \to L^{5/4}$. Theorem 2.8 applies and yields boundedness of T_m in the entire region $L^p \times L^q \to L^r$ with $1 < p, q < \infty$ and $1/2 < r < \infty$. $\qquad\square$

We end this section by noting that the Marcinkiewicz multiplier theorem does not hold for bilinear operators; see [16].

2.6 The bilinear Hilbert transform and the method of rotations

It is also natural to ask whether Theorem 2.12 is true under less stringent conditions on the function Ω; for instance, instead of assuming that Ω is a Lipschitz function on the sphere, can one assume that it is in some L^q for $q > 1$? It is a classical result obtained by Calderón and Zygmund [4] using the method of rotations that homogeneous linear singular integrals with odd kernels are always L^p bounded for $1 < p < \infty$. We indicate what happens if the method of rotations is used in the multilinear setting. Let Ω be an odd integrable function on $\mathbf{S}^{2n-1}$. Let

$$T_\Omega(f_1, f_2)(x) = \iint_{\mathbf{R}^{2n}} \frac{\Omega((y_1, y_2)/|(y_1, y_2)|)}{|(y_1, y_2)|^{2n}} \, f_1(x - y_1) \, f_2(x - y_2) \, dy_1 \, dy_2.$$

Using polar coordinates in $\mathbf{R}^{2n}$, we express

$$T_\Omega(f_1, f_2)(x) = \int_{\mathbf{S}^{2n-1}} \Omega(\theta_1, \theta_2) \left\{ \int_0^{+\infty} f_1(x - t\theta_1) \, f_2(x - t\theta_2) \, \frac{dt}{t} \right\} d(\theta_1, \theta_2).$$

Replacing (θ_1, θ_2) by $-(\theta_1, \theta_2)$, changing variables, and using that Ω is odd, we obtain

$$T_\Omega(f_1, f_2)(x) = \int_{\mathbf{S}^{2n-1}} \Omega(\theta_1, \theta_2) \left\{ \int_0^{+\infty} f_1(x + t\theta_1) \, f_2(x + t\theta_2) \, \frac{dt}{t} \right\} d(\theta_1, \theta_2),$$

and averaging these identities we deduce that

$$T_\Omega(f_1, f_2)(x) = \frac{1}{2} \int_{\mathbf{S}^{2n-1}} \Omega(\theta_1, \theta_2) \left\{ \int_{-\infty}^{+\infty} f_1(x - t\theta_1) \, f_2(x - t\theta_2) \, \frac{dt}{t} \right\} d(\theta_1, \theta_2).$$

The method of rotations gives rise to the operator inside the curly brackets above and one would like to know that this operator is bounded from a product of two Lebesgue spaces into another Lebesgue space (and preferably) uniformly bounded in θ_1, θ_2. Motivated by this calculation, for vectors $u, v \in \mathbf{R}^n$ we introduce the family of operators

$$\mathcal{H}_{u,v}(f_1, f_2)(x) = \text{p.v.} \int_{-\infty}^{+\infty} f_1(x - tu) \, f_2(x - tv) \, \frac{dt}{t}.$$

We call this operator the *directional bilinear Hilbert transform* (in the direction indicated by the vector (u, v) in $\mathbf{R}^{2n}$). In the special case $n = 1$, we use the notation

$$H_{\alpha,\beta}(f, g)(x) = \text{p.v.} \int_{-\infty}^{+\infty} f(x - \alpha t) \, g(x - \beta t) \, \frac{dt}{t}$$

for the bilinear Hilbert transform defined for functions f, g on the line and for $x, \alpha, \beta \in \mathbf{R}$.

We mention results concerning boundedness of these operators. The operator $H_{\alpha,\beta}$ was first shown to be bounded by Lacey and Thiele [28], [29] in the range

$$1 < p, q \le \infty, \qquad 2/3 < r < \infty, \qquad 1/p + 1/q = 1/r. \tag{2.37}$$

Uniform L^r bounds (in α, β) for $H_{\alpha,\beta}$ were obtained by Grafakos and Li [17] in the local L^2 case (i.e., the case when $2 < p, q, r' < \infty$) and extended by Li [32] in the hexagonal region

$$1 < p, q, r < \infty, \qquad \left| \frac{1}{p} - \frac{1}{q} \right| < \frac{1}{2}, \qquad \left| \frac{1}{p} - \frac{1}{r'} \right| < \frac{1}{2}, \qquad \left| \frac{1}{q} - \frac{1}{r'} \right| < \frac{1}{2}. \tag{2.38}$$

The bilinear Hilbert transforms first appeared in an attempt of Calderón to show that the first commutator

$$C_1(f; A)(x) = \text{p.v.} \int_{\mathbf{R}} \left(\frac{A(x) - A(y)}{x - y} \right) \frac{f(y)}{x - y} \, dy$$

is L^2 bounded. In fact, in the mid 1960's Calderón observed that the commutator $f \mapsto C_1(f; A)$ can be written as the average

$$C_1(f; A)(x) = \int_0^1 H_{1,\alpha}(f, A')(x) \, d\alpha,$$

and the boundedness of $C_1(f; A)$ can therefore be reduced to the uniform (in α) boundedness of $H_{1,\alpha}$. Likewise, the uniform boundedness of the $H_{\alpha,\beta}$ can be used to show that T_Ω is bounded from $L^{p_1}(\mathbf{R}) \times L^{p_2}(\mathbf{R})$ to $L^p(\mathbf{R})$ when Ω is an odd function on the sphere $\mathbf{S}^1$.

We use a similar idea to obtain new bounds for a higher-dimensional commutator introduced by Christ and Journé [5]. The n-dimensional commutator is defined as

$$\mathcal{C}_1^{(n)}(f,a)(x) = \text{p.v.} \int_{\mathbf{R}^n} K(x-y) \int_0^1 f(y)\, a((1-t)x + ty)\, dt\, dy, \qquad (2.39)$$

where $K(x)$ is a Calderón–Zygmund kernel in dimension n, and f, a are functions on $\mathbf{R}^n$. Christ and Journé [5] proved that $\mathcal{C}_1^{(n)}$ is bounded from $L^p(\mathbf{R}^n) \times L^\infty(\mathbf{R}^n)$ to $L^p(\mathbf{R}^n)$ for $1 < p < \infty$. Here we discuss some off-diagonal bounds $L^p \times L^q \to L^r$ whenever $1/p + 1/q = 1/r$ and $1 < p, q, r < \infty$.

As the operator $\mathcal{C}_1^{(n)}(f,a)$ is n-dimensional, we will need to "transfer" $H_{\alpha,\beta}$ in higher dimensions. To achieve this we use rotations. We have the following lemma:

Lemma 2.16. *Suppose that K is a kernel in $\mathbf{R}^{2n}$ (which may be a distribution) and let T_K be the bilinear singular integral operator associated with K,*

$$T_K(f,g)(x) = \iint K(x-y, x-z)\, f(y)\, g(z)\, dy\, dz.$$

Assume that T_K is bounded from $L^p(\mathbf{R}^n) \times L^q(\mathbf{R}^n) \to L^r(\mathbf{R}^n)$ with norm $\|T\|$ when $1/p + 1/q = 1/r$. Let M be a $n \times n$ invertible matrix. Define a $2n \times 2n$ invertible matrix

$$\widetilde{M} = \begin{pmatrix} M & O \\ O & M \end{pmatrix},$$

where O is the zero $n \times n$ matrix. Then the operator $T_{K \circ \widetilde{M}}$ is also bounded from $L^p(\mathbf{R}^n) \times L^q(\mathbf{R}^n) \to L^r(\mathbf{R}^n)$ with norm at most $\|T\|$.

Proof. To prove the lemma we note that

$$T_{K \circ \widetilde{M}}(f,g)(x) = T_K(f \circ M^{-1}, g \circ M^{-1})(Mx),$$

from which it follows that

$$\begin{aligned}
\left\| T_{K \circ \widetilde{M}}(f,g) \right\|_{L^r} &= (\det M)^{-1/r} \left\| T_K(f \circ M^{-1}, g \circ M^{-1}) \right\|_{L^r} \\
&\leq (\det M)^{-1/r} \|T\| \, \|f \circ M^{-1}\|_{L^p} \|g \circ M^{-1}\|_{L^q} \\
&= \|T\| \, (\det M)^{-1/r} \|T\| \, \|f\|_{L^p} (\det M)^{1/p} \|g\|_{L^q} (\det M)^{1/p} \\
&= \|T\| \, \|f\|_{L^p} \|g\|_{L^q}. \qquad \qquad \square
\end{aligned}$$

We apply Lemma 2.16 to the bilinear Hilbert transform. Let $e_1 = (1, 0, \ldots, 0)$ be the standard coordinate vector on $\mathbf{R}^n$. We begin with the observation that the operator $\mathcal{H}_{\alpha e_1, \beta e_1}(f,g)$ defined for functions f, g on $\mathbf{R}^n$ is bounded from $L^p(\mathbf{R}^n) \times L^q(\mathbf{R}^n)$ to $L^r(\mathbf{R}^n)$ for the same range of indices as the bilinear Hilbert transform. Indeed, the operator $\mathcal{H}_{\alpha e_1, \beta e_1}$ can be viewed as the classical one-dimensional bilinear Hilbert transform in the coordinate x_1 followed by the identity operator in the remaining coordinates $x_2, \ldots, x_n$, where $x = (x_1, \ldots, x_n)$. By Lemma 2.16, for an invertible $n \times n$ matrix M and $x \in \mathbf{R}^n$,

$$\mathcal{H}_{\alpha e_1, \beta e_1}(f \circ M^{-1}, g \circ M^{-1})(Mx) = \text{p.v.} \int_{-\infty}^{+\infty} f(x - \alpha t M^{-1} e_1) \, g(x - \beta t M^{-1} e_1) \, \frac{dt}{t}$$

maps $L^p(\mathbf{R}^n) \times L^q(\mathbf{R}^n) \to L^r(\mathbf{R}^n)$ with norm the same as the one-dimensional bilinear Hilbert transform $H_{\alpha, \beta}$ whenever the indices p, q, r satisfy (2.37). If M is a rotation (i.e., an orthogonal matrix), then $M^{-1} e_1$ can be any unit vector in $\mathbf{S}^{n-1}$. We conclude that the family of operators

$$\mathcal{H}_{\alpha\theta, \beta\theta}(f, g)(x) = \text{p.v.} \int_{-\infty}^{+\infty} f(x - \alpha t \, \theta) \, g(x - \beta t \, \theta) \, \frac{dt}{t}, \qquad x \in \mathbf{R}^n$$

is bounded from $L^p(\mathbf{R}^n) \times L^q(\mathbf{R}^n)$ to $L^r(\mathbf{R}^n)$ with a bound independent of $\theta \in \mathbf{S}^{n-1}$ whenever the indices p, q, r satisfy (2.37). This bound is also independent of α, β whenever the indices p, q, r satisfy (2.38).

It remains to express the higher-dimensional commutator $\mathcal{C}_1^{(n)}$ in terms of the operators $\mathcal{H}_{\alpha\theta, \beta\theta}$. Here we make the assumption that K is an odd homogeneous singular integral operator on $\mathbf{R}^n$, such as a Riesz transform. For a fixed $x \in \mathbf{R}^n$ we apply polar coordinates centered at x by writing $y = x - r\theta$. Then we can express the higher-dimensional commutator in (2.39) as

$$\int_{\mathbf{S}^{n-1}} \int_0^\infty \frac{K(\theta)}{r^n} \int_0^1 f(x - r\theta) \, a(x - tr\theta) \, dt \, r^{n-1} \, dr \, d\theta. \tag{2.40}$$

Changing variables from $\theta \to -\theta$, $r \to -r$, and using that $K(\theta)$ is odd, we write this expression as

$$\int_{\mathbf{S}^{n-1}} \int_{-\infty}^0 K(\theta) \int_0^1 f(x - r\theta) \, a(x - tr\theta) \, dt \, \frac{dr}{r} \, d\theta. \tag{2.41}$$

Averaging (2.40) and (2.41) we arrive at the identity

$$\mathcal{C}_1^{(n)}(f, a)(x) = \frac{1}{2} \int_{\mathbf{S}^{n-1}} K(\theta) \int_0^1 \mathcal{H}_{\theta, t\theta}(f, a)(x) \, dt \, d\theta.$$

This identity implies boundedness of $\mathcal{C}_1^{(n)}$ from $L^p(\mathbf{R}^n) \times L^q(\mathbf{R}^n)$ to $L^r(\mathbf{R}^n)$ whenever the indices p, q, r satisfy (2.38). Interpolation with the known $L^p \times L^\infty \to L^p$ bounds yields the following:

Theorem 2.17. *Let K be an odd homogeneous singular integral on $\mathbf{R}^n$. Then the n-dimensional commutator $\mathcal{C}_1^{(n)}$ associated with K maps $L^p(\mathbf{R}^n) \times L^q(\mathbf{R}^n)$ to $L^r(\mathbf{R}^n)$ whenever $1/p + 1/q = 1/r$ and $(1/p, 1/q, 1/r)$ lies in the open convex hull of the pentagon with vertices $(0, 1/2, 1/2)$, $(0, 0, 0)$, $(1, 0, 1)$, $(1/2, 1/2, 1)$, and $(1/6, 4/6, 5/6)$.*

2.7 Counterexample for the higher-dimensional bilinear ball multiplier

In this section we address the question whether the bilinear multiplier operator with symbol the characteristic function of the unit ball B in $\mathbf{R}^{2n}$,

$$T_{\chi_B}(f,g)(x) = \int_{\mathbf{R}^n} \int_{\mathbf{R}^n} \widehat{f}(\xi)\, \widehat{g}(\eta)\, \chi_{|\xi|^2+|\eta|^2<1}\, e^{2\pi i x \cdot (\xi+\eta)}\, d\xi\, d\eta, \qquad (2.42)$$

is a bounded bilinear operator from $L^p(\mathbf{R}^n) \times L^q(\mathbf{R}^n)$ to $L^r(\mathbf{R}^n)$ for some indices p, q, r related as in Hölder's inequality. We adapt Fefferman's counterexample [13] for the ball multiplier on L^p, $p \neq 2$ to the bilinear setting for indices outside the local L^2 case. We consider this problem only in dimension 2, since it can be extended to higher dimensions via a bilinear adaptation of de Leeuw's theorem [30], proved in [11]. The results in this section can be found in [11].

For a rectangle R in $\mathbf{R}^2$, let R' be the union of the two copies of R adjacent to R in the direction of its longest side. Hence, $R \cup R'$ is a rectangle three times as long as R with the same center. Key to this argument is the following geometric lemma, whose proof can be found in [14] or [42].

Lemma 2.18. *Let $\delta > 0$ be given. Then there exists a measurable subset E of $\mathbf{R}^2$ and a finite collection of rectangles R_j in $\mathbf{R}^2$ such that:*

(1) *The R_j are pairwise disjoint.*

(2) *We have $1/2 \leq |E| \leq 3/2$.*

(3) *We have $|E| \leq \delta \sum_j |R_j|$.*

(4) *For all j we have $|R_j' \cap E| \geq \frac{1}{12}|R_j|$.*

Let $\delta > 0$ and let E and R_j be as in Lemma 2.18. The proof of Lemma 2.18 implies that there are 2^k rectangles R_j of dimension $2^{-k} \times 3\log(k+2)$. Here, k is chosen so that $k+2 \geq e^{1/\delta}$. Let v_j be the unit vector in $\mathbf{R}^2$ parallel to the longest side of R_j and in the direction of the set E indicated by the longest side of R_j.

Lemma 2.19. *Let R be a rectangle in $\mathbf{R}^2$ and let v be a unit vector in $\mathbf{R}^2$ parallel to the longest side of R. Let R' be as above. Consider the half-space $\mathcal{H}_v$ of $\mathbf{R}^4$ defined by*

$$\mathcal{H}_v = \{(\xi,\eta) \in \mathbf{R}^2 \times \mathbf{R}^2 : (\xi+\eta) \cdot v \geq 0\}.$$

Then the following estimate is valid for all $x \in \mathbf{R}^2$:

$$\left| \int_{\mathbf{R}^2} \int_{\mathbf{R}^2} \chi_{\mathcal{H}_v}(\xi,\eta)\, \widehat{\chi_R}(\xi)\, \widehat{\chi_R}(\eta)\, e^{2\pi i x \cdot (\xi+\eta)}\, d\xi\, d\eta \right| \geq \frac{1}{10}\chi_{R'}(x). \qquad (2.43)$$

Proof. We introduce an orthogonal matrix $\mathcal{O}$ of $\mathbf{R}^2$ such that $\mathcal{O}(v) = (1,0)$. Setting $\xi = (\xi_1, \xi_2)$, $\eta = (\eta_1, \eta_2)$, we write the expression on the left in (2.43) as

$$\left| \iint_{\mathcal{O}^{-1}(\xi+\eta)\cdot v \geq 0} \widehat{\chi_R}(\mathcal{O}^{-1}\xi)\, \widehat{\chi_R}(\mathcal{O}^{-1}\eta)\, e^{2\pi i x \cdot \mathcal{O}^{-1}(\xi+\eta)}\, d\xi\, d\eta \right|$$

$$= \left| \iint_{\xi_1 + \eta_1 \geq 0} \widehat{\chi_{\mathcal{O}[R]}}(\xi)\, \widehat{\chi_{\mathcal{O}[R]}}(\eta)\, e^{2\pi i \mathcal{O}x \cdot (\xi+\eta)}\, d\xi\, d\eta \right|.$$

Now the rectangle $\mathcal{O}[R]$ has sides parallel to the axes, say $\mathcal{O}[R] = I_1 \times I_2$. Assume that $|I_1| > |I_2|$, i.e., its longest side is horizontal. Let H be the classical Hilbert transform on the line. Setting $\mathcal{O}x = (y_1, y_2)$, we can write the last displayed expression as

$$\left| \chi_{I_2}(y_2)^2 \int_{\xi_1 \in \mathbf{R}} \widehat{\chi_{I_1}}(\xi_1)\, e^{2\pi i y_1 \xi_1} \int_{\eta_1 \geq -\xi_1} \widehat{\chi_{I_1}}(\eta_1)\, e^{2\pi i y_1 \eta_1}\, d\eta_1\, d\xi_1 \right|$$

$$= \chi_{I_2}(y_2) \left| \int_{\xi_1 \in \mathbf{R}} \widehat{\chi_{I_1}}(\xi_1)\, \tfrac{1}{2}(I + iH)\big[\chi_{I_1}(\,\cdot\,)\, e^{2\pi i \xi_1(\,\cdot\,)}\big](y_1)\, d\xi_1 \right|$$

$$= \chi_{I_2}(y_2) \left| \tfrac{1}{2}(I + iH)\, \chi_{I_1}(y_1) \right|$$

$$= \left| \big[\chi_{\xi_1 \geq 0}\, \widehat{\chi_{I_1 \times I_2}}(\xi_1, \xi_2)\big]^{\vee}(y_1, y_2) \right|.$$

Using the result from [14, Proposition 10.1.2] or [42, estimate (33), page 453], we deduce that the preceding expression is at least

$$\frac{1}{10} \chi_{(I_1 \times I_2)'}(y_1, y_2) = \frac{1}{10} \chi_{(\mathcal{O}[R])'}(\mathcal{O}x) = \frac{1}{10} \chi_{R'}(x).$$

This proves the required conclusion.   □

Lemma 2.20. *Let* $v_1, v_2, \ldots, v_j, \ldots$ *be a sequence of unit vectors in* $\mathbf{R}^2$. *Define a sequence of half-spaces* $\mathcal{H}_{v_j}$ *in* $\mathbf{R}^4$ *as in Lemma 2.19. Let* B, B_1, B_2 *be the following sets in* $\mathbf{R}^4$:

$$B = \{(\xi, \eta) \in \mathbf{R}^2 \times \mathbf{R}^2 \,:\, |\xi|^2 + |\eta|^2 \leq 1\}$$

$$B^{*1} = \{(\xi, \eta) \in \mathbf{R}^2 \times \mathbf{R}^2 \,:\, |\xi + \eta|^2 + |\eta|^2 \leq 1\}$$

$$B^{*2} = \{(\xi, \eta) \in \mathbf{R}^2 \times \mathbf{R}^2 \,:\, |\xi|^2 + |\xi + \eta|^2 \leq 1\}.$$

Assume that one of T_{χ_B}, $T_{\chi_{B*1}}$, $T_{\chi_{B*2}}$ is bounded from $L^p(\mathbf{R}^2) \times L^q(\mathbf{R}^2)$ to $L^r(\mathbf{R}^2)$ and has norm $C = C(p, q, r)$. Then the vector-valued inequality

$$\left\| \left(\sum_j |T_{\chi_{\mathcal{H}_{v_j}}}(f_j, g_j)|^2 \right)^{1/2} \right\|_{L^r(\mathbf{R}^2)}$$

$$\leq C \left\| \left(\sum_j |f_j|^2 \right)^{1/2} \right\|_{L^p(\mathbf{R}^2)} \left\| \left(\sum_j |g_j|^2 \right)^{1/2} \right\|_{L^q(\mathbf{R}^2)}$$

holds for all functions f_j and g_j.

Proof. Assume that T_{χ_B} is bounded from $L^p(\mathbf{R}^2) \times L^q(\mathbf{R}^2)$ to $L^r(\mathbf{R}^2)$ for some indices $p, q, r > 0$. Set $\xi = (\xi_1, \xi_2)$ and $\eta = (\eta_1, \eta_2) \in \mathbf{R}^2$. For $\rho > 0$ we define sets

$$B_\rho = \{(\xi, \eta) \in \mathbf{R}^2 \times \mathbf{R}^2 \ : \ |\xi|^2 + |\eta|^2 \leq 2\rho^2\}$$

$$B_{j,\rho} = \{(\xi, \eta) \in \mathbf{R}^2 \times \mathbf{R}^2 \ : \ |\xi - \rho v_j|^2 + |\eta - \rho v_j|^2 \leq 2\rho^2\}.$$

Note that bilinear multiplier norms are translation and dilation invariant; consequently, we have

$$\|T_{\chi_{B_{j,\rho}}}\|_{L^p \times L^q \to L^r} = \|T_{\chi_{B_\rho}}\|_{L^p \times L^q \to L^r} = \|T_{\chi_B}\|_{L^p \times L^q \to L^r} = C. \tag{2.44}$$

Moreover, in view of the bilinear version of a theorem of Marcinkiewicz and Zygmund ([19, §9]), we have the following vector-valued extension of our boundedness assumption on $T_{\chi_{B_\rho}}$:

$$\left\| \left(\sum_j |T_{\chi_{B_\rho}}(f_j, g_j)|^2 \right)^{1/2} \right\|_{L^r} \leq C \left\| \left(\sum_j |f_j|^2 \right)^{1/2} \right\|_{L^p} \left\| \left(\sum_j |g_j|^2 \right)^{1/2} \right\|_{L^q}.$$

Since $\chi_{B_{j,\rho}} \to \chi_{\mathcal{H}_{v_j}}$ pointwise as $\rho \to \infty$ for $x \in \mathbf{R}^2$, we deduce that

$$\lim_{\rho \to \infty} T_{\chi_{B_{j,\rho}}}(f, g)(x) = T_{\chi_{\mathcal{H}_{v_j}}}(f, g)(x)$$

for suitable functions f and g. By Fatou's lemma we conclude that

$$\left\| \left(\sum_j |T_{\chi_{\mathcal{H}_{v_j}}}(f_j, g_j)|^2 \right)^{1/2} \right\|_{L^r} \leq \liminf_{\rho \to \infty} \left\| \left(\sum_j |T_{\chi_{B_{j,\rho}}}(f_j, g_j)|^2 \right)^{1/2} \right\|_{L^r}. \tag{2.45}$$

Now, observe the following identity:

$$T_{\chi_{B_{j,\rho}}}(f, g)(x) = e^{4\pi i \rho v_j \cdot x} T_{\chi_{B_\rho}}(e^{-2\pi i \rho v_j \cdot (\cdot)} f, \, e^{-2\pi i \rho v_j \cdot (\cdot)} g)(x). \tag{2.46}$$

Using (2.45) and (2.46), we obtain

$$\left\|\left(\sum_j |T_{\chi_{\mathcal{H}_j}}(f_j, g_j)|^2\right)^{1/2}\right\|_{L^r}$$

$$\leq \liminf_{\rho\to\infty} \left\|\left(\sum_j |e^{4\pi i\rho v_j\cdot(\cdot)} T_{\chi_{B_\rho}}(e^{-2\pi i\rho v_j\cdot(\cdot)}f_j, \, e^{-2\pi i\rho v_j\cdot(\cdot)}g_j)|^2\right)^{1/2}\right\|_{L^r}$$

$$\leq \liminf_{\rho\to\infty} \|T_{\chi_{B_\rho}}\|_{L^p\times L^q\to L^r} \left\|\left(\sum_j |e^{-2\pi i\rho v_j\cdot(\cdot)}f_j|^2\right)^{1/2}\right\|_{L^p}$$

$$\left\|\left(\sum_j |e^{-2\pi i\rho v_j\cdot(\cdot)}g_j|^2\right)^{1/2}\right\|_{L^q}$$

$$= C \left\|\left(\sum_j |f_j|^2\right)^{1/2}\right\|_{L^p} \left\|\left(\sum_j |g_j|^2\right)^{1/2}\right\|_{L^q},$$

where in the last step we used (2.44).

The proof of the analogous statements for $T_{B^{*1}}$ and $T_{B^{*2}}$ is as follows. We introduce sets

$$B_\rho^{*1} = \{(\xi, \eta) \in \mathbf{R}^2 \times \mathbf{R}^2 : |\xi + \eta|^2 + |\eta|^2 \leq \rho^2\}$$

$$B_{j,\rho}^{*1} = \{(\xi, \eta) \in \mathbf{R}^2 \times \mathbf{R}^2 : |\xi - \rho v_j + \eta|^2 + |\eta|^2 \leq \rho^2\}$$

$$B_\rho^{*2} = \{(\xi, \eta) \in \mathbf{R}^2 \times \mathbf{R}^2 : |\xi|^2 + |\xi + \eta|^2 \leq \rho^2\}$$

$$B_{j,\rho}^{*2} = \{(\xi, \eta) \in \mathbf{R}^2 \times \mathbf{R}^2 : |\xi|^2 + |\xi + \eta - \rho v_j|^2 \leq \rho^2\}$$

and we note that the characteristic functions of $B_{j,\rho}^{*1}$ and $B_{j,\rho}^{*2}$ converge to the characteristic function of $\mathcal{H}_{v_j}$ as $\rho \to \infty$. Using the identities

$$T_{\chi_{B_{j,\rho}^{*1}}}(f, g)(x) = e^{2\pi i\rho v_j\cdot x} T_{\chi_{B_\rho^{*1}}}(e^{-2\pi i\rho v_j\cdot(\cdot)}f, g)(x),$$

$$T_{\chi_{B_{j,\rho}^{*2}}}(f, g)(x) = e^{2\pi i\rho v_j\cdot x} T_{\chi_{B_\rho^{*2}}}(f, e^{-2\pi i\rho v_j\cdot(\cdot)}g)(x),$$

we obtain the same conclusion assuming boundedness of the bilinear operators $T_{\chi_{B^{*1}}}$ and $T_{\chi_{B^{*2}}}$ from $L^p(\mathbf{R}^2) \times L^q(\mathbf{R}^2)$ to $L^r(\mathbf{R}^2)$. $\quad\square$

The following is the main result of this section.

Theorem 2.21. *Fix indices p, q, r satisfying $0 < p, q, r < \infty$ and $1/p + 1/q = 1/r$ in the non-local L^2 region, i.e., in the region where one of p, q, or r' is less than 2. Let B be the unit ball in $\mathbf{R}^4$. Then the operator in (2.42) (with $n = 2$) is not bounded from $L^p(\mathbf{R}^2) \times L^q(\mathbf{R}^2)$ to $L^r(\mathbf{R}^2)$.*

Proof. First fix p, q, r satisfying $p^{-1} + q^{-1} = r^{-1} < 1/2$ with $r > 2$. To obtain a contradiction, we assume that the operator in (2.42) (with $n = 2$) is bounded from $L^p(\mathbf{R}^2) \times L^q(\mathbf{R}^2)$ to $L^r(\mathbf{R}^2)$ with norm C.

Suppose that $\delta > 0$ is given. Let E and R_j be as in Lemma 2.18. Let v_j be the unit vector parallel to the longest side of R_j and pointing in the direction of the set E indicated by the longest side of R_j. We have

$$\sum_j \int_E \left| T_{\mathcal{H}_{v_j}}(\chi_{R_j}, \chi_{R_j})(x) \right|^2 dx$$

$$\leq |E|^{\frac{r-2}{r}} \left\| \left(\sum_j |T_{\mathcal{H}_{v_j}}(\chi_{R_j}, \chi_{R_j})|^2 \right)^{1/2} \right\|_{L^r}^2$$

$$\leq C |E|^{\frac{r-2}{r}} \left\| \left(\sum_j |\chi_{R_j}|^2 \right)^{1/2} \right\|_{L^p}^2 \left\| \left(\sum_j |\chi_{R_j}|^2 \right)^{1/2} \right\|_{L^q}^2$$

$$= C |E|^{\frac{r-2}{r}} \left(\sum_j |R_j| \right)^{2/r} \leq C \delta^{\frac{r-2}{r}} \sum_j |R_j|,$$

where we used Hölder's inequality with $r > 2$, Lemma 2.20, the disjointness of the rectangles R_j, and Lemma 2.18, respectively, in the preceding sequence of estimates.

We also have a reverse inequality:

$$\sum_j \int_E \left| T_{\mathcal{H}_{v_j}}(\chi_{R_j}, \chi_{R_j})(x) \right|^2 dx \geq \sum_j \int_E \left(\tfrac{1}{10} \chi_{R_j'}(x) \right)^2 dx$$

$$= \tfrac{1}{100} \sum_j |E \cap R_j'| \geq \tfrac{1}{1200} \sum_j |R_j|,$$

where we used Lemma 2.19 and Lemma 2.18.

Combining the upper and lower estimates for $\sum_j \int_E \left| T_{\mathcal{H}_{v_j}}(\chi_{R_j}, \chi_{R_j})(x) \right|^2 dx$, we obtain the inequality

$$\tfrac{1}{1200} \sum_j |R_j| \leq C \delta^{\frac{r-2}{r}} \sum_j |R_j|,$$

and therefore

$$\tfrac{1}{1200} \leq C \delta^{\frac{r-2}{r}}$$

for any $\delta > 0$. This is a contradiction since we are assuming that $r > 2$.

The lack of boundedness of the ball multiplier operator (2.42) in the remaining non-local L^2 regions ($p > 2$, $q < 2$, $r < 2$) and ($p < 2$, $q > 2$, $r < 2$) follows by duality, while in the region ($1 \leq p, q < \infty$, $\frac{1}{2} < r \leq 1$) it is obvious. $\quad\square$

Bibliography

[1] A. Benedek, A. Calderón, and R. Panzone, *Convolution operators on Banach-space valued functions*, Proc. Nat. Acad. Sci. USA **48** (1962), 356–365.

[2] G. Bourdaud, *Une algèbre maximale d'opérateurs pseudo-différentiels*, Comm. Partial Differential Equations **13** (1988), 1059–1083.

[3] A. Calderón and A. Zygmund, *On the existence of certain singular integrals*, Acta Math. **88** (1952), 85–139.

[4] A. Calderón and A. Zygmund, *On singular integrals*, Amer. J. Math. **78** (1956), 289–309.

[5] M. Christ and J.-L. Journé, *Polynomial growth estimates for multilinear singular integral operators*, Acta Math. **159** (1987), 51–80.

[6] R. R. Coifman and Y. Meyer, *On commutators of singular integrals and bilinear singular integrals*, Trans. Amer. Math. Soc. **212** (1975), 315–331.

[7] R. R. Coifman and Y. Meyer, *Commutateurs d'intégrales singulières et opérateurs multilinéaires*, Ann. Inst. Fourier (Grenoble) **28** (1978), 177–202.

[8] R. R. Coifman and Y. Meyer, *Au-delà des opérateurs pseudo-différentiels*, Astérisque **57**, Société Mathématique de France, Paris, 1978.

[9] R. R. Coifman and Y. Meyer, *Non-linear harmonic analysis, operator theory, and PDE*, in: Beijing Lectures in Harmonic Analysis, E. M. Stein, ed., Ann. of Math. Studies **112**, Princeton University Press, Princeton, NJ, 1986.

[10] G. David and J.-L. Journé, *A boundedness criterion for generalized Calderón–Zygmund operators*, Ann. of Math. **120** (1984), 371–397.

[11] G. Diestel and L. Grafakos, *Unboundedness of the ball bilinear multiplier operator*, Nagoya Math. J. **185** (2007), 151–159.

[12] G. Diestel and N. Kalton, personal communication.

[13] C. Fefferman, *The multiplier problem for the ball*, Ann. of Math. **94** (1971), 330–336.

[14] L. Grafakos, *Modern Fourier Analysis*, 2nd edition, Graduate Texts in Math. **250**, Springer, New York, 2008.

[15] L. Grafakos and N. Kalton, *Some remarks on multilinear maps and interpolation*, Math. Ann. **319** (2001), 151–180.

[16] L. Grafakos and N. Kalton, *The Marcinkiewicz multiplier condition for bilinear operators*, Studia Math. **146** (2001), 115–156.

[17] L. Grafakos and X. Li, *Uniform bounds for the bilinear Hilbert transforms I*, Ann. of Math. **159** (2004), 889–933.

[18] L. Grafakos and X. Li, *The disc as a bilinear multiplier*, Amer. J. Math. **128** (2006), 91–119.

[19] L. Grafakos and J. M. Martell, *Extrapolation of weighted norm inequalities for multivariable operators*, J. Geom. Anal. **14** (2004), 19–46.

[20] L. Grafakos and J. Soria, *Translation-invariant bilinear operators with positive kernels*, Integral Equations Operator Theory **66** (2010), 253–264.

[21] L. Grafakos and T. Tao, *Multilinear interpolation between adjoint operators*, J. Funct. Anal. **199** (2003), 379–385.

[22] L. Grafakos and R. H. Torres, *Multilinear Calderón–Zygmund theory*, Adv. Math. **165** (2002), 124–164.

[23] L. Hörmander, *Estimates for translation invariant operators in L^p spaces*, Acta Math. **104** (1960), 93–140.

[24] L. Hörmander, *The Analysis of Linear Partial Differential Operators I*, 2nd edition, Springer, Berlin Heidelberg New York, 1990.

[25] S. Janson, *On interpolation of multilinear operators*, in: Function Spaces and Applications (Lund, 1986), Lecture Notes in Math. **1302**, Springer, Berlin Heidelberg New York, 1988.

[26] J.-L. Journé, *Calderón–Zygmund operators on product spaces*, Rev. Mat. Iberoamericana **1** (1985), 55–91.

[27] C. Kenig and E. M. Stein, *Multilinear estimates and fractional integration*, Math. Res. Lett. **6** (1999), 1–15.

[28] M. T. Lacey and C. M. Thiele, *L^p bounds for the bilinear Hilbert transform, $2 < p < \infty$*, Ann. of Math. **146** (1997), 693–724.

[29] M. T. Lacey and C. M. Thiele, *On Calderón's conjecture*, Ann. of Math. **149** (1999), 475–496.

[30] K. de Leeuw, *On L_p multipliers*, Ann. of Math. **81** (1965), 364–379.

[31] A. Lerner, S. Ombrosi, C. Pérez, R. Torres, and R. Trujillo-González, *New maximal functions and multiple weights for the multilinear Calderón–Zygmund theory*, Adv. Math. **220** (2009), 1222–1264.

[32] X. Li, *Uniform bounds for the bilinear Hilbert transforms II*, Rev. Mat. Iberoamericana **22** (2006), 1069–1126.

[33] Y. Meyer and R. R. Coifman, *Wavelets: Calderón–Zygmund and Multilinear Operators*, Cambridge University Press, Cambridge, UK, 1997.

[34] S. G. Mihlin, *On the multipliers of Fourier integrals* [in Russian], Dokl. Akad. Nauk. **109** (1956), 701–703.

[35] J. Peetre, *On convolution operators leaving $L^{p,\lambda}$ spaces invariant*, Ann. Mat. Pura Appl. **72** (1966), 295–304.

[36] F. Rubio de Francia, *A Littlewood–Paley inequality for arbitrary intervals*, Rev. Mat. Iberoamericana **1** (1985), 1–14.

[37] S. Sato, *Note on Littlewood–Paley operator in higher dimensions*, J. London Math. Soc. **42** (1990), 527–534.

[38] F. Soria, *A note on a Littlewood–Paley inequality for arbitrary intervals in $\mathbb{R}^2$*, J. London Math. Soc. **36** (1987), 137–142.

[39] S. Spanne, *Sur l'interpolation entre les espaces* $\mathcal{L}_k^{p\Phi}$, Ann. Scuola Norm. Sup. Pisa **20** (1966), 625–648.

[40] E. M. Stein, *Singular integrals, harmonic functions, and differentiability properties of functions of several variables*, in: Singular Integrals, Proc. Sympos. Pure Math. **10** (1967), 316–335.

[41] E. M. Stein, *Singular Integrals and Differentiability Properties of Functions*, Princeton University Press, Princeton, NJ, 1970.

[42] E. M. Stein, *Harmonic Analysis: Real Variable Methods, Orthogonality, and Oscillatory Integrals*, Princeton University Press, Princeton, NJ, 1993.

[43] R. Strichartz, *A multilinear version of the Marcinkiewicz interpolation theorem*, Proc. Amer. Math. Soc. **21** (1969), 441–444.

[44] C. Thiele, *A uniform estimate*, Ann. of Math. **157** (2002), 1–45.

Chapter 3

Singular Integrals and Weights

Carlos Pérez[1]

Summary

This chapter is an expanded version of the material covered in a minicourse given at the Centre de Recerca Matemàtica in Barcelona during the week May 4–8, 2009. We provide details and different proofs of known results as well as new ones. We also survey on several recent results related to the core of this course, namely weighted optimal bounds for Calderón–Zygmund operators with weights. The basic topics covered by the lectures revolved around the Rubio de Francia iteration algorithm, the extrapolation theorem with optimal bounds, the Coifman–Fefferman estimate, the Besicovitch covering lemma, and rearrangements of functions. These notes can be seen as a modern introduction to the A_p theory of weights.

3.1   Introduction

The *Hardy–Littlewood maximal function* is the operator defined by

$$Mf(x) = \sup_{x \in Q} \frac{1}{|Q|} \int_Q |f(y)| \, dy,$$

[1]The author would like to thank Professors Joan Mateu and Joan Orobitg for coordinating two special minicourses at the Centre de Recerca Matemàtica on Multilinear Harmonic Analysis and Weights presented by the author and Professor Loukas Grafakos during the period May 4–8, 2009. These minicourses were part of a special research program for the academic year 2008–2009 entitled *Harmonic Analysis, Geometric Measure Theory, and Quasiconformal Mappings*, coordinated by Professors Xavier Tolsa and Joan Verdera. The author would also like to thank the Centre de Recerca Matemàtica for the invitation to spend the semester there and to give this course. The author acknowledges support from the Spanish Ministry of Science and Innovation through grant MTM2009-08934 and from Junta de Andalucía through the *proyecto de excelencia* FQM-4745.

where the supremum is taken over all the cubes containing x and where f is any locally integrable function. It is clear that M is not a linear operator but it is a sort of self-dual operator, since the following inequality holds:

$$\sup_{\lambda>0} \lambda w(\{x \in \mathbb{R}^n \ : \ Mf(x) > \lambda\}) \leq c \int_{\mathbb{R}^n} fMw\,dx, \tag{3.1}$$

for all nonnegative functions f and w. Here it is crucial that the constant c is independent of both functions f and w. We use the standard notation $w(E) = \int_E w(x)\,dx$, where E is any measurable set.

The inequality (3.1) is interesting on its own because it is an improvement of the classical weak-type $(1,1)$ property of the Hardy–Littlewood maximal operator M. However, the crucial new point of view is that it can be seen as a sort of duality for M, since the following L^p inequality holds:

$$\int_{\mathbb{R}^n} (Mf)^p w\,dx \leq c_p \int_{\mathbb{R}^n} |f|^p Mw\,dx, \qquad f, w \geq 0. \tag{3.2}$$

This estimate follows from the classical interpolation theorem of Marcinkiewicz. Both (3.1) and (3.2) were proved by C. Fefferman and E. M. Stein in [29] to derive the following vector-valued extension of the classical Hardy–Littlewood maximal theorem: For all $1 < p, q < \infty$ there is a finite constant $c = c_{p,q}$ such that

$$\left\| \left(\sum_j (Mf_j)^q \right)^{\frac{1}{q}} \right\|_{L^p(\mathbb{R}^n)} \leq c \left\| \left(\sum_j |f_j|^q \right)^{\frac{1}{q}} \right\|_{L^p(\mathbb{R}^n)}. \tag{3.3}$$

This is a very deep theorem and has been used a lot in modern harmonic analysis explaining the central role of (3.1). Nevertheless, the proof of (3.1) does not follow from the classical maximal theorem (corresponding to the case $w \equiv 1$), but the proof is nearly identical and is based on a covering lemma of Vitali type, as can be seen for instance in [27] or [31]. We show in Section 3.2 a simpler and direct proof based on the classical Besicovitch covering lemma.

The Muckenhoupt–Wheeden conjecture

In these lectures we are mainly interested in corresponding estimates for Calderón–Zygmund operators T instead of M. Here we use the standard concept of Calderón–Zygmund operator as can be found in many places, as for instance in [33].

Conjecture 3.1 (Muckenhoupt–Wheeden conjecture)*. There exists a constant c such that, for any function f and any weight w,*

$$\sup_{\lambda>0} \lambda w(\{x \in \mathbb{R}^n \ : \ |Tf(x)| > \lambda\}) \leq c \int_{\mathbb{R}^n} |f|\, Mw\,dx. \tag{3.4}$$

The author was studying this problem during the 90's and only much later he found out that it had been studied by B. Muckenhoupt and R. Wheeden during the

70's. In particular, it seems that these authors conjectured that (3.4) should hold for $T = H$, the simplest singular integral operator, namely the Hilbert transform:

$$Hf(x) = \text{p.v.} \int_{\mathbb{R}} \frac{f(y)}{x-y}\, dy.$$

We could think that to obtain a vector-valued extension of the classical Calderón–Zygmund theorem for singular integral operators similar to (3.3), namely

$$\left\| \left(\sum_j |Tf_j|^q \right)^{\frac{1}{q}} \right\|_{L^p(\mathbb{R}^n)} \leq c \left\| \left(\sum_j |f_j|^q \right)^{\frac{1}{q}} \right\|_{L^p(\mathbb{R}^n)},$$

we would need to prove (3.4). However, less refined estimates such as

$$\sup_{\lambda>0} \lambda w(\{x \in \mathbb{R}^n : |Tf(x)| > \lambda\}) \leq c_r \int_{\mathbb{R}^n} |f|\, M_r w\, dx$$

would do the job, where $M_r w = (M(w^r))^{1/r}$, $r > 1$. This was shown by A. Córdoba and C. Fefferman in [17], being the key fact that $M_r w \in A_1$ if $r > 1$ (the Coifman–Rochberg estimate (3.42)). This result shows that inequalities with weights having some additional properties yield interesting results as well.

It seems that the best result obtained is due to the author and can be found in [64], where M is replaced by $M_{L(\log L)^\epsilon}$, an "ϵ-logarithmically" bigger maximal type operator than M. If $\epsilon = 0$ we recover M, but the constant c_ϵ blows up as $\epsilon \to 0$. The precise result is the following.

Theorem 3.2 (The $L(\log L)^\epsilon$ theorem). *There exists a constant c depending on T such that, for any $\epsilon > 0$, any function f and any weight w,*

$$\sup_{\lambda>0} \lambda w(\{x \in \mathbb{R}^n : |Tf(x)| > \lambda\}) \leq \frac{c}{\epsilon} \int_{\mathbb{R}^n} |f|\, M_{L(\log L)^\epsilon} w\, dx, \qquad w \geq 0. \quad (3.5)$$

We remark that the operator $M_{L(\log L)^\epsilon}$ is pointwise smaller than M_r, $r > 1$.

Remark 3.3. Very recently and in a joint work with T. Hytönen we have improved this theorem, replacing T by the maximal singular integral operator T_*. The main difficulty is that T_* is not linear and the approach used in these notes cannot be applied; see [39].

The author conjectured in [64] that the inequality (3.4) would be false. This was confirmed in [9] in the case of fractional integrals I_α which are positive operators. However, there were some evidences suggesting that the conjecture could be true. Indeed, the conjecture was confirmed by Chanillo and Wheeden in [11] for certain continuous Littlewood–Paley square function S (see (3.23)) instead of H, namely

$$\sup_{t>0} tw(\{x \in \mathbb{R}^n : Sf(x) > t\}) \leq c \int_{\mathbb{R}^n} |f|\, Mw\, dx, \qquad w \geq 0. \quad (3.6)$$

Also a similar result was obtained in [67] for the vector-valued maximal function (see (3.3)):

$$\sup_{t>0} tw\left(\left\{x \in \mathbb{R}^n \ : \ \left(\sum_{i=1}^\infty (Mf_i(x))^q\right)^{1/q} > t\right\}\right) \leq c_q \int_{\mathbb{R}^n} |f(x)|_q \, Mw(x) \, dx, \quad (3.7)$$

where $w \geq 0$ and $1 < q < \infty$. Since many evidences show that the vector-valued maximal function $\left(\sum_{i=1}^\infty (Mf_i(x))^q\right)^{1/q}$ behaves somehow like a singular integral, both inequalities (3.6) and (3.7) suggested that it would be the same for the case of singular integrals.

The Muckenhoupt–Wheeden conjecture remained open even for the Hilbert transform until the end of 2010, when it was disproved for the Hilbert transform by M. C. Reguera and C. Thiele [75]. Previously, Reguera had shown that (3.4) is false for dyadic type singular operators T, more precisely for a special Haar multiplier operator. This was the main result of Reguera's Ph.D. thesis [73]. Haar multipliers can be seen as dyadic versions of singular integrals and are used as models to understand them. It is remarkable that Reguera disproved a stronger weighted L^2 result of special type, namely of the form

$$\int_{\mathbb{R}} |Tf|^2 \, w \, dx \leq c \int_{\mathbb{R}} |f|^2 \left(\frac{Mw}{w}\right)^2 w \, dx, \qquad w \geq 0. \quad (3.8)$$

Indeed, it was shown in [23] that, if (3.4) holds for an arbitrary operator T, then this weighted L^2 estimate holds for T. In other words, Reguera disproved a stronger inequality than the original one of Muckenhoupt–Wheeden, at least for these special dyadic singular integrals. This result gave strong evidence for a negative answer of the Muckenhoupt–Wheeden conjecture.

This scheme was used in the subsequent paper [75] by Reguera and Thiele, where they gave a simplified construction of the weight given in [74] and finally showed that (3.4) is really false for the Hilbert transform H, by showing again that (3.8) is false for H. This result is really interesting and is related to what is called the A_1 conjecture, that we discuss below and which is really the main motivation of these lecture notes.

The A_1 conjecture

This is a variant of Conjecture 3.1 and the idea is to assume an a priori condition on the weight w. This condition can be read directly from the Fefferman–Stein inequality (3.1) and in fact it was already introduced by these authors in that paper: the weight w is an A_1 weight or satisfies the A_1 condition if there is a finite constant c such that

$$Mw \leq cw \quad \text{a.e.} \quad (3.9)$$

It is standard to denote by $[w]_{A_1}$ the smallest of these constants c. Then, if $w \in A_1$,

$$\sup_{t>0} tw(\{x \in \mathbb{R}^n \ : \ Mf(x) > t\}) \leq c_n \, [w]_{A_1} \int_{\mathbb{R}^n} |f| \, w \, dx, \quad (3.10)$$

and it is natural to ask whether the corresponding inequality holds for singular integrals (say for the Hilbert transform):

Conjecture 3.4 (A_1 conjecture). *If $w \in A_1$, then*

$$\sup_{\lambda > 0} \lambda w(\{x \in \mathbb{R}^n : |Tf(x)| > \lambda\}) \leq c\,[w]_{A_1} \int_{\mathbb{R}^n} |f|\, w\, dx. \tag{3.11}$$

However, this inequality seems to be false too (see [58]) for $T = H$, the Hilbert transform.

In this lecture we survey on some recent progress in connection with this conjecture and exhibit an extra logarithmic growth in (3.11), which, in view of [58], could be the best possible result. To prove this logarithmic growth result, we have to study first the corresponding weighted $L^p(w)$ estimates with $1 < p < \infty$ and $w \in A_1$, being the result this time fully sharp. The final parts of the proofs of both theorems can be found in Sections 3.6 and 3.7 and are essentially taken from [54] and [52].

Strong type estimates

To study inequality (3.11), it is natural to ask first the dependence of $\|T\|_{L^p(w)}$, $p > 1$, in terms of $[w]_{A_1}$. We discuss briefly some results before the papers [52, 54].

Denote by δ the best possible exponent in the inequality

$$\|T\|_{L^p(w)} \leq c_{n,p}\,[w]_{A_1}^{\delta}. \tag{3.12}$$

In the case when $p = 2$ and $T = H$ is the Hilbert transform, R. Fefferman and J. Pipher [28] established that $\delta = 1$. The proof is based on sharp A_1 bounds for appropriate square functions on $L^2(w)$ from the works [10, 11]. In particular, the following celebrated inequality of Chang–Wilson–Wolff was used:

$$\int_{\mathbb{R}^n} (Sf)^2 w\, dx \leq C \int_{\mathbb{R}^n} |f|^2 Mw\, dx,$$

where S is any of the classical Littlewood–Paley square functions, as for instance (3.23) (compare with inequality (3.6)). One can show that this approach yields $\delta = 1$ also for $p > 2$. However, for $1 < p < 2$ the same approach gives the estimate $\delta \leq 1/2 + 1/p$. Also, that approach works only for smooth singular integrals of convolution type, and recall that Calderón–Zygmund operators are non-convolution operators with a very minimal regularity condition.

In [52] and [54] a different approach was used to show that, for any Calderón–Zygmund operator, the sharp exponent in (3.12) is $\delta = 1$ for all $1 < p < \infty$. The method has its roots in the classical Calderón–Zygmund theory but with several extra refinements. We believe that the circle of ideas in these papers may lead to a proof of the A_2 conjecture 3.12 below.

We state now our main theorems. From now on, T will always denote any Calderón–Zygmund operator and we assume that the reader is familiar with the classical unweighted theory.

Theorem 3.5 (Linear growth theorem). *Let T be a Calderón–Zygmund operator and let $1 < p < \infty$. Then*

$$\|T\|_{L^p(w)} \leq cpp'\,[w]_{A_1},\tag{3.13}$$

where $c = c(n, T)$.

As an application of this result, we obtain the following endpoint estimate.

Theorem 3.6 (Logarithmic growth theorem). *Let T be a Calderón–Zygmund operator. Then*

$$\|T\|_{L^{1,\infty}(w)} \leq c\,[w]_{A_1}(1 + \log[w]_{A_1}),\tag{3.14}$$

where $c = c(n, T)$.

Remark 3.7. The result in (3.13) is best possible and, as already mentioned, [58] strongly suggests that (3.14) could also be the best possible result. On the other hand, and very recently, a new improvement of these two theorems has been found by the author and T. Hytönen [38] in terms of mixed A_1-A_∞ constants. See Section 3.9 for some details, in particular Theorem 3.33 and also Theorem 3.14.

Remark 3.8. As in Remark 3.3, these two theorems can be further improved by replacing T by the maximal singular integral operator T_*. Again, the method presented in these notes cannot be applied because it is based on the fact that T is linear while T_* is not; see [39].

The weak (p, p) conjecture and Rubio de Francia's algorithm

If we could improve (3.14) by removing the log term, that is, if the A_1 conjecture were to hold, then we would have the following result.

Conjecture 3.9 (Weak (p, p) conjecture). *Let $1 < p < \infty$ and let T be a Calderón–Zygmund singular integral operator. There is a constant $c = c(n, T)$ such that, for any A_p weight w,*

$$\|T\|_{L^{p,\infty}(w)} \leq cp\,[w]_{A_p}.\tag{3.15}$$

We recall that a weight w satisfies the *Muckenhoupt A_p condition* if

$$[w]_{A_p} \equiv \sup_Q \left(\frac{1}{|Q|}\int_Q w(x)\,dx\right)\left(\frac{1}{|Q|}\int_Q w(x)^{-1/(p-1)}\,dx\right)^{p-1} < \infty.$$

Here $[w]_{A_p}$ is usually called the A_p *constant* (also called *characteristic* or *norm*) of the weight. The case $p = 1$ is understood by replacing the right-hand side by $(\inf_Q w)^{-1}$, which is equivalent to the definition given above in (3.9).

Observe the duality relationship:

$$[w]_{A_p}^{1/p} = [\sigma]_{A_{p'}}^{1/p'},$$

where we use the standard notation $\sigma = w^{\frac{-1}{p-1}} = w^{1-p'}$. Note also that $[w]_{A_p} \geq 1$.

In Section 3.4.1 we will prove this conjecture assuming that the A_1 conjecture is true. The argument will be based on an application of the Rubio de Francia algorithm or scheme. The same argument applied to the inequality (3.14) yields the following result.

Corollary 3.10. *Let $1 < p < \infty$ and let T be a Calderón–Zygmund operator. If $w \in A_p$, then*

$$\|Tf\|_{L^{p,\infty}(w)} \le cp\,[w]_{A_p}(1 + \log[w]_{A_p})\|f\|_{L^p(w)}, \tag{3.16}$$

where $c = c(n, p, T)$.

Observe that, for p close to one, the behavior of the constant is much better than in (3.21). The advantage here is that this method works for any Calderón–Zygmund operator.

Rubio de Francia's algorithm is a technique which has become ubiquitous in the modern theory of weights and beyond. It is also very flexible, as will be shown in Sections 3.3 and 3.4, where it will be applied to five different scenarios of interest for these lecture notes. We refer the reader to the monograph [20] for a full account of the technique.

The A_2 conjecture

In [56], Muckenhoupt proved the fundamental result characterizing all the weights for which the Hardy–Littlewood maximal operator is bounded on $L^p(w)$; the surprisingly simple necessary and sufficient condition is the celebrated A_p condition of Muckenhoupt. Of course, the operator norm $\|M\|_{L^p(w)}$ will depend on the A_p condition of w, but it seems that the first precise result was proved by S. Buckley [8] as part of his Ph.D. thesis.

Theorem 3.11. *If $w \in A_p$, then the Hardy–Littlewood maximal function satisfies the following operator estimate:*

$$\|M\|_{L^p(w)} \le c_n p' \, [w]_{A_p}^{\frac{1}{p-1}},$$

or, equivalently,

$$\sup_{w \in A_p} \frac{1}{[w]_{A_p}^{\frac{1}{p-1}}} \|M\|_{L^p(w)} \le c_n p'. \tag{3.17}$$

Furthermore, the result is sharp in the sense that, for any $\theta > 0$,

$$\sup_{w \in A_p} \frac{1}{[w]_{A_p}^{\frac{1}{p-1} - \theta}} \|M\|_{L^p(w)} = \infty. \tag{3.18}$$

In fact, we cannot replace the function $\psi(t) = t^{\frac{1}{p-1}}$ by a "smaller" function $\psi \colon [1, \infty) \to (0, \infty)$ in the sense that

$$\inf_{t>1} \frac{\psi(t)}{t^{\frac{1}{p-1}}} = 0$$

(or $\lim_{t\to\infty} \frac{\psi(t)}{t^\beta} = 0$, or $\sup_{t>1} \frac{t^\beta}{\psi(t)} = \infty$, or $\lim_{t\to\infty} \frac{t^\beta}{\psi(t)} = \infty$), since then

$$\sup_{w\in A_p} \frac{1}{\psi([w]_{A_p})} \|M\|_{L^p(w)} = \infty. \tag{3.19}$$

The original proof of Buckley is delicate because it is based on a sharp version of the so-called Reverse Hölder Inequality for A_p weights. However, very recently, A. Lerner [48] has found a very nice and simple proof of this result that will be given in Section 3.2. It is based on the Besicovitch lemma, which can be avoided when M is a dyadic maximal function. To see the power of this new proof, we remark that the constant c_p in Buckley's proof cannot be so precise. In fact, in these lecture notes we avoid the use of the Besicovitch lemma by considering first the dyadic case and finally "shifting" the dyadic cubes.

In fact, (3.17) should be compared with the weak-type bound

$$\|M\|_{L^p(w)\to L^{p,\infty}(w)} \leq c_n [w]_{A_p}^{1/p}, \tag{3.20}$$

whose proof is much simpler and will be shown in Lemma 3.17.

Buckley's theorem attracted renewed attention after the work of Astala, Iwaniec and Saksman [3] on the theory of quasiregular mappings. They proved sharp regularity results for solutions to the Beltrami equation, assuming that the operator norm of the Beurling–Ahlfors transform grows linearly in terms of the A_p constant for $p \geq 2$. This linear growth was proved by Petermichl and Volberg in [72]. This result opened up the possibility of considering some other operators such as the classical Hilbert transform. Finally, Petermichl [70, 71] proved the corresponding results for the Hilbert transform and the Riesz transforms. To be more precise, it has been shown in [70, 71, 72] that if T is either the Beurling–Ahlfors, Hilbert or Riesz transform and $1 < p < \infty$, then

$$\|T\|_{L^p(w)} \leq c_{n,p} [w]_{A_p}^{\max\left\{1, \frac{1}{p-1}\right\}}. \tag{3.21}$$

Furthermore, the exponent $\max\left\{1, \frac{1}{p-1}\right\}$ is best possible by examples similar to the one related to Theorem 3.11.

This result should be compared with the linear growth theorem (Theorem 3.5). Indeed, recall that $A_1 \subset A_p$ and that

$$[w]_{A_p} \leq [w]_{A_1}.$$

Therefore, (3.21) clearly gives that $\delta = 1$ in (3.12) when $p \geq 2$. However, (3.21) cannot be used in Theorem 3.5 to get the sharp exponent δ in the range $1 < p < 2$, becoming the exponent worst when p gets close to 1.

In view of these results and others (for instance the case of paraproducts [6], due to O. Beznosova), it was believed that the conjecture that should be true is the following.

Conjecture 3.12 (A_2 conjecture). *Let $1 < p < \infty$ and let T be a Calderón–Zygmund singular integral operator. Then there is a constant $c = c(n, T)$ such that, for any A_p weight w,*

$$\|T\|_{L^p(w)} \leq cpp' \, [w]_{A_p}^{\max\left\{1, \frac{1}{p-1}\right\}}. \tag{3.22}$$

This conjecture has been proved by Hytönen in [36]. We will briefly mention in the next paragraphs some of the previous steps done toward this conjecture, although the main topic of these lectures is more related to the A_1 conjecture already mentioned.

The maximum in the exponent reflects the duality of T, namely that T^* is also a Calderón–Zygmund operator. In fact, it can be shown that if T is self-adjoint (or essentially like Calderón–Zygmund operators) and if (3.22) is proved for $p > 2$ then the case $1 < p < 2$ follows by duality (see Corollary 3.20). What is more interesting is that, by the sharp Rubio de Francia extrapolation theorem obtained in [26] (or by Corollary 3.20 again), it is enough to prove (3.22) only for $p = 2$. This is the reason why the A_p result has been called the A_2 conjecture. Observe that in this case the growth of the constant is simply linear. In these lecture notes we prove a special case of the extrapolation theorem, which is enough for our purposes, namely Theorem 3.31 and Corollary 3.20.

In each of the previously known cases, the proof of (3.22) – again, just in the case $p = 2$ by the sharp Rubio de Francia extrapolation theorem – is based on a technique developed by Petermichl [70] reducing the problem to proving the analogous inequality for a corresponding Haar shift operator. The norm inequalities for these dyadic operators were then proved using Bellman function techniques. Much more recently, Lacey, Petermichl and Reguera–Rodríguez [45] gave a proof of the A_2 conjecture for a large family of Haar shift operators that includes all the dyadic operators needed for the above results. Their proof avoids the use of Bellman functions, and instead uses a deep, two-weight testing type "Tb theorem" for Haar shift operators due to Nazarov, Treil and Volberg [59]. A bit later and motivated by [45], a completely different proof was found by the author together with D. Cruz-Uribe and J. M. Martell in [22], which avoids both Bellman functions and two-weight norm inequalities such as the Tb theorem. Instead, we used a very interesting and flexible decomposition formula for general functions f due to Lerner [50], whose main ideas go back to the work of Fuji [30]. The main new ingredient is the use of local mean oscillation instead of the usual oscillation, which was the one considered by Fuji.

We remit the reader to [22] for details of how to apply Lerner's formula to generalized Haar shift operators. An important advantage of this approach (again by means of Lerner's formula) is that it also yields the optimal sharp one-weight norm inequalities for other operators such as dyadic square functions and paraproducts for the vector-valued maximal function of Fefferman–Stein, as well as some very sharp two-weight "A_p bump" type conditions. All these results can be found in [21].

As a sample, we mention the following result for dyadic square functions. Let Δ denote the collection of dyadic cubes in $\mathbb{R}^n$. Given $Q \in \Delta$, let $\widehat{Q}$ be its dyadic parent: the unique dyadic cube containing Q whose side-length is twice that of Q. The dyadic square function is the operator

$$ S_d f(x) = \left(\sum_{Q \in \Delta} (f_Q - f_{\widehat{Q}})^2 \chi_Q(x) \right)^{1/2}, \tag{3.23} $$

where $f_Q = \frac{1}{|Q|} \int_Q f$. For the properties of the dyadic square function, we refer the reader to Wilson [80].

Theorem 3.13. *Given p with $1 < p < \infty$, we have, for any $w \in A_p$,*

$$ \|S_d f\|_{L^p(w)} \leq C_{n,p} [w]_{A_p}^{\max\left\{ \frac{1}{2}, \frac{1}{p-1} \right\}} \|f\|_{L^p(w)}. $$

Furthermore, the exponent $\max\left\{ \frac{1}{2}, \frac{1}{p-1} \right\}$ is the best possible.

The exponent in Theorem 3.13 was first conjectured by Lerner [47] for the continuous square function; he also showed that it was the best possible. An interesting fact concerning the proof of this theorem is that its proof is based again on the sharp Rubio de Francia extrapolation theorem. The novelty is that the extrapolation hypothesis is for the case $p = 3$ instead of $p = 2$ as in the case of singular integrals. Again, details can be found in [21].

The results mentioned above for generalized Haar shifts could be used to prove the A_2 conjecture when the kernel of the Calderón–Zygmund operator is sufficiently smooth (for instance, C^2 would be enough, by applying for instance the approximating result of Vagharshakyan [77]). However, this was not enough to prove the full A_2 conjecture, since it is assumed that the kernel satisfies merely a Hölder–Lipschitz condition. Finally, the conjecture was proved, as already mentioned, by T. Hytönen in August 2010 [36]. The proof was based on an important reduction obtained by the author with S. Treil and A. Volberg in [69]. Very roughly, this reduction says that a weighted L^2 weak type estimate is essentially equivalent to prove the corresponding strong type. A bit later, a direct proof, avoiding this reduction, was found in [40]. One of the key points is to use a probabilistic representation formula due to Hytönen. Then the generalized shift operators act as building blocks of this representation.

Therefore, an important and hard part of the proof of the A_2 conjecture was to obtain bounds for appropriate Haar shift operators with "complexity (m, n)" that depends at most polynomially on the complexity – the problem in [21] is that the method is very flexible but the complexity's dependence is of exponential type. The estimate obtained in [40] is of polynomial degree $k = 3$ that was further improved to linear by [43] and [76]. The last result was based on the Bellman method using some ideas from another argument given in [60] with a slightly worst estimate but with the advantage that it can be transferred to the context of doubling metric spaces.

Improving the A_2 conjecture, now theorem

On the other hand, and in a direction that we think is more interesting, the A_2 conjecture, which is now a theorem, has been improved by the author and Hytönen [38] in terms of mixed A_2-A_∞ constants (see remark 3.7), and in [37] for a general p. We state this new result, whose proof is based on a new sharp reverse Hölder property for A_∞ weights (Theorem 3.47), that can also be found in [38]. The new idea is to derive results fractioning the A_2 constant in two fractions, one involving the A_2 constant as such and the other involving the A_∞ constant.

Theorem 3.14. *Let T be a Calderón–Zygmund operator. Then there is a constant c depending on T such that*

$$\|T\|_{L^2(w)} \le c\, [w]_{A_2}^{1/2} \, \max\{[w]_{A_\infty}, [w^{-1}]_{A_\infty}\}^{1/2}.$$

See Section 3.9 for some details of these mixed A_p-A_∞ constants mainly in the case $p = 1$.

The "fractional" A_2 conjecture

We finish this introductory section by mentioning briefly some results about fractional integrals that confirm both Conjecture 3.9 and Conjecture 3.12. Of course these operators are different from singular integrals, but this is still a good indication.

Recall that, for $0 < \alpha < n$, the fractional integral operator or Riesz potential I_α is defined, except perhaps for a constant, by

$$I_\alpha f(x) = \int_{\mathbb{R}^n} \frac{f(y)}{|x-y|^{n-\alpha}}\, dy.$$

In [57], Muckenhoupt and Wheeden characterized the weighted strong-type inequality for fractional operators in terms of the so-called $A_{p,q}$ condition. For $1 < p < \frac{n}{\alpha}$ and q defined by $\frac{1}{q} = \frac{1}{p} - \frac{\alpha}{n}$, they showed that

$$\left(\int_{\mathbb{R}^n} (w I_\alpha f)^q \, dx \right)^{1/q} \le c \left(\int_{\mathbb{R}^n} (w f)^p \, dx \right)^{1/p}, \qquad f \ge 0 \qquad (3.24)$$

if and only if $w \in A_{p,q}$:

$$[w]_{A_{p,q}} \equiv \sup_Q \left(\frac{1}{|Q|} \int_Q w^q \, dx \right) \left(\frac{1}{|Q|} \int_Q w^{-p'} \, dx \right)^{q/p'} < \infty.$$

In [44] the following estimates were shown. Suppose that p, q, α are as above. Then:

- (The weak estimate)

$$\|I_\alpha f\|_{L^{q,\infty}(w^q)} \leq c\, [w]_{A_{p,q}}^{1-\frac{\alpha}{n}} \|wf\|_{L^p(\mathbb{R}^n)}, \qquad f \geq 0.$$

- (The strong estimate)

$$\|wI_\alpha f\|_{L^{q,\infty}(\mathbb{R}^n)} \leq c\, [w]_{A_{p,q}}^{\left(1-\frac{\alpha}{n}\right)\max\left\{1,\frac{p'}{q}\right\}} \|wf\|_{L^p(\mathbb{R}^n)}, \qquad f \geq 0.$$

Furthermore, both exponents are sharp.

Note that, if we formally put $\alpha = 0$ in these results, we may think that the fractional integral becomes a singular integral operator and we recover the two conjectures already mentioned.

3.2 Three applications of the Besicovitch covering lemma to the maximal function

In this section we take the maximal function as a model example to understand more difficult operators. Before considering the strong case (Theorem 3.11), we will show a corresponding result for the weak type case. To prove both theorems, we will use the classical Besicovitch covering lemma.

Lemma 3.15 (Besicovitch covering lemma). *Let K be a bounded set in $\mathbb{R}^n$ and suppose that for every $x \in K$ there is an (open) cube $Q(x)$ with center at x. Then we can find a (possibly finite) sequence of points $\{x_j\}$ in K such that*

$$K \subset \bigcup_j Q(x_j) \qquad and \qquad \sum_j \chi_{Q(x_j)} \leq c_n,$$

where c_n is a finite-dimensional constant.

The proof of this result can be found in several places, such as the classical lecture notes by M. de Guzmán [25], or also in [33].

We distinguish two cases, namely $p = 1$ and $1 < p < \infty$. The first case will follow after proving the Fefferman–Stein basic initial estimate (3.1).

Lemma 3.16. *There is a dimensional constant c such that, for all f and w,*

$$\sup_{\lambda>0} \lambda w(\{x \in \mathbb{R}^n : Mf(x) > \lambda\}) \leq c \int_{\mathbb{R}^n} |f(x)|\, Mw(x)\, dx \qquad (3.25)$$

and hence

$$\|M\|_{L^{1,\infty}(w)} \leq c_n\, [w]_{A_1}.$$

Proof. The proof is just an application of the Besicovitch covering lemma. Indeed, assuming that w is bounded as we may, the first observation is that (3.25) is equivalent to

$$\sup_{\lambda>0} \lambda w(\{x \in \mathbb{R}^n : M\left(f\frac{w}{Mw}\right)(x) > \lambda\}) \le C \int_{\mathbb{R}^n} |f(x)|\, w(x)\, dx, \qquad \lambda > 0.$$

The second is that we trivially have the pointwise inequality

$$M\left(f\frac{w}{Mw}\right)(x) \le c_n M_w^c f(x),$$

where M_w^c is the weighted centered maximal function

$$M_w^c f(x) = \sup_{r>0} \frac{1}{w(Q_r(x))} \int_{Q_r(x)} |f(y)|\, w(y)\, dy. \qquad (3.26)$$

Therefore (3.25) follows from

$$\sup_{\lambda>0} \lambda w(\{x \in \mathbb{R}^n : M_w^c f(x) > \lambda\}) \le C \int_{\mathbb{R}^n} |f(x)|\, w(x)\, dx,$$

which is a consequence of the Besicovitch covering lemma, where C is a dimensional constant. $\qquad\square$

Lemma 3.17. *Let $w \in A_p$, $1 < p < \infty$. There is a constant $c = c_n$ such that, for any A_p weight w,*

$$\|M\|_{L^{p,\infty}(w)} \le c_n\, [w]_{A_p}^{1/p}. \qquad (3.27)$$

This result is known but the point we want to make is to compare the exponent of the A_p constant with the other exponents appearing in these lecture notes. For instance, compare it with both exponents in Conjecture 3.9 and Theorem 3.5.

Another interesting observation here is that, if we consider the dual estimate, the constant is essentially the same, namely

$$\|M\|_{L^{p',\infty}(\sigma)} \le c_n\, [\sigma]_{A_{p'}}^{1/p'} = c_n\, [w]_{A_p}^{1/p}, \qquad (3.28)$$

where we recall that $\sigma = w^{1-p'}$.

Proof of the Lemma. Since $w \in A_p$ for each cube Q and nonnegative function f,

$$\left(\frac{1}{|Q|} \int_Q f(y)\, dy\right)^p w(Q) \le [w]_{A_p} \int_Q f(y)^p w(y)\, dy$$

and hence

$$Mf(x) \approx M^c f(x) \le [w]_{A_p}^{1/p} M_w^c (f^p)(x)^{1/p}.$$

We finish by applying again the Besicovitch covering lemma:

$$\|Mf\|_{L^{p,\infty}(w)} \leq c_n\, [w]_{A_p}^{1/p} \|M_w^c(f^p)^{1/p}\|_{L^{p,\infty}(w)} \leq c_n\, [w]_{A_p}^{1/p} \|M_w^c(f^p)\|_{L^{1,\infty}(w)}^{1/p}$$

$$\leq c_n\, [w]_{A_p}^{1/p} \left(\int_{\mathbb{R}^n} f^p w\, dx \right)^{1/p}. \qquad \qquad \square$$

We remark that there are other proofs of these lemmas without appealing to the Besicovitch lemma, just by a Vitali type covery lemma. We leave the proof to the interested reader.

We now prove Buckley's Theorem 3.11 based on Lerner's proof [48]. This proof has as a bonus an improvement of the constant $c_{n,p}$.

Proof of Theorem 3.11. To prove (3.17) we set, for any cube Q,

$$A_p(Q) = \frac{w(Q)}{|Q|} \left(\frac{\sigma(Q)}{|Q|} \right)^{p-1},$$

where, as usual, $\sigma = w^{1-p'}$. Now we will first consider the case of the dyadic maximal function:

$$\frac{1}{|Q|} \int_Q |f| = A_p(Q)^{\frac{1}{p-1}} \left\{ \frac{|Q|}{w(Q)} \left(\frac{1}{\sigma(Q)} \int_Q |f| \right)^{p-1} \right\}^{\frac{1}{p-1}}$$

$$\leq [w]_{A_p}^{\frac{1}{p-1}} \left\{ \frac{1}{w(Q)} \int_Q M_\sigma^d(f\sigma^{-1})^{p-1}\, dx \right\}^{\frac{1}{p-1}},$$

where M_σ^d is the weighted dyadic maximal function. Hence,

$$M^d f(x) \leq [w]_{A_p}^{\frac{1}{p-1}} \left\{ M_w^d \big(M_\sigma^d(f\sigma^{-1})^{p-1} w^{-1} \big)(x) \right\}^{\frac{1}{p-1}}.$$

We conclude by using that

$$\|M_\mu^d\|_{L^p(\mu)} \leq p'$$

with bounds independent of μ, which follows from the improved version of the Marcinkewicz interpolation theorem in the following form:

$$\|T\|_{L^p(\mu)} \leq p'\, \|T\|_{L^{1,\infty}(\mu)}^{1/p} \|T\|_{L^\infty(\mu)}^{1/p'}, \qquad 1 < p < \infty,$$

where T is any sublinear operator bounded on $L^\infty(\mu)$ and of weak type $(1,1)$ with norms $\|T\|_{L^\infty(\mu)}$ and $\|T\|_{L^{1,\infty}(\mu)}$ respectively (see for instance [32, p. 42, Exercise 1.3.3]). This gives the estimate

$$\|M^d\|_{L^p(w)} \leq [w]_{A_p}^{\frac{1}{p-1}}\, p^{p'/p}\, p' \leq [w]_{A_p}^{\frac{1}{p-1}}\, e\, p',$$

finishing the proof in the case of the dyadic maximal function. Observe that the A_p constant here is the dyadic A_p constant. The general situation follows easily by "shifting" the dyadic network applying Minkowski's inequality to the following well-known Fefferman–Stein shifting lemma that can be found in [31, p. 431]: For each integer k,

$$M^{2^k} f(x) \leq \frac{2^{3n+1}}{|Q_{2^{k+2}}(0)|} \int_{Q_{2^{k+2}}(0)} \left(\tau_{-t} \circ M^d \circ \tau_t\right) f(x)\, dt, \qquad x \in \mathbb{R}^n,$$

where $\tau_t g(x) = g(x-t)$, $Q_r(0)$ is the cube centered at the origin with side length r, and M^δ, $\delta > 0$ is the operator defined as M but with cubes having side length smaller than δ.

For the sharpness we consider $n = 1$ and $0 < \epsilon < 1$. Let

$$w(x) = |x|^{(1-\epsilon)(p-1)}.$$

It is easy to check that

$$[w]_{A_p}^{\frac{1}{p-1}} \approx \frac{1}{\epsilon}.$$

Let also

$$f(y) = y^{-1+\epsilon(p-1)} \chi_{(0,1)}(y)$$

and observe that

$$\|f\|_{L^p(w)}^p \approx \frac{1}{\epsilon}.$$

To estimate now $\|Mf\|_{L^p(w)}$, we pick $0 < x < 1$; thus,

$$Mf(x) \geq \frac{1}{x} \int_0^x f(y)\, dy = c_p \frac{1}{\epsilon} f(x)$$

and hence

$$\|Mf\|_{L^p(w)} \geq c_p \frac{1}{\epsilon} \|f\|_{L^p(w)},$$

from which the sharpness (3.18) follows easily. $\qquad\square$

3.3 Two applications of Rubio de Francia's algorithm: Optimal factorization and extrapolation

3.3.1 The sharp factorization theorem

Muckenhoupt already observed in [56] that it follows from the definition of the A_1 class of weights that if $w_1, w_2 \in A_1$, then the weight

$$w = w_1 w_2^{1-p}$$

is an A_p weight. Furthermore we have

$$[w]_{A_p} \leq [w_1]_{A_1} [w_2]_{A_1}^{p-1}. \tag{3.29}$$

He conjectured that any A_p weight can be written in this way. This conjecture was proved by P. Jones, namely if $w \in A_p$ then there are A_1 weights w_1, w_2 such that $w = w_1 w_2^{1-p}$.

It is also well known that the modern approach to this question uses a completely different path and it is due to J. L. Rubio de Francia, as can be found in [33], where we remit the reader for more information about the A_p theory of weights. Here we present a variation which appears in [34] and we give a proof of this result using sharp constants. To be more precise, we have the following result.

Lemma 3.18. *Let $1 < p < \infty$ and let $w \in A_p$. Then there are A_1 weights $u, v \in A_1$ such that*

$$w = u\, v^{1-p}$$

in such a way that

$$[u]_{A_1} \leq c_n [w]_{A_p} \quad and \quad [v]_{A_1} \leq c_n [w]_{A_p}^{\frac{1}{p-1}}, \tag{3.30}$$

and hence $[w]_{A_p} \leq [u]_{A_1} [v]_{A_1}^{p-1} \leq c_n [w]_{A_p}^2$.

Proof. We use Rubio de Francia's iteration scheme or algorithm to our situation. Define

$$S_1(f)^{p'} \equiv w^{1/p} M\left(\frac{|f|^{p'}}{w^{1/p}}\right)$$

and

$$S_2(f)^p \equiv \frac{1}{w^{1/p}} M(|f|^p w^{1/p}).$$

Observe that $S_i \colon L^{pp'}(\mathbb{R}^n) \to L^{pp'}(\mathbb{R}^n)$ with constant

$$\|S_i\|_{L^{pp'}(\mathbb{R}^n)} \leq c_n [w]_{A_p}^{1/p}, \qquad i = 1, 2,$$

by Buckley's theorem.

Now, the operator $S = S_1 + S_2$ is bounded on $L^{pp'}(\mathbb{R}^n)$ with

$$\|S\|_{L^{pp'}(\mathbb{R}^n)} \leq c_n [w]_{A_p}^{1/p}.$$

Define the Rubio de Francia algorithm R by

$$R(h) \equiv \sum_{k=0}^{\infty} \frac{1}{2^k} \frac{S^k(h)}{\left(\|S\|_{L^{pp'}(\mathbb{R}^n)}\right)^k}.$$

Observe that R is also bounded on $L^{pp'}(\mathbb{R}^n)$. Now, if $h \in L^{pp'}(\mathbb{R}^n)$ is fixed, then $R(h) \in A_1(S)$. More precisely,

$$S(R(h)) \le 2\,\|S\|_{L^{pp'}(\mathbb{R}^n)} \le c_n\,[w]_{A_p}^{1/p}.$$

In particular, $R(h) \in A_1(S_i)$, $i = 1, 2$, with

$$S_i(R(h)) \le c_n\,[w]_{A_p}^{1/p} R(h), \qquad i = 1, 2.$$

Hence,

$$M(R(h)^{p'}w^{-1/p}) \le c_n\,[w]_{A_p}^{p'/p} R(h)^{p'}w^{-1/p}$$

and

$$M(R(h)^p w^{1/p}) \le c_n\,[w]_{A_p} R(h)^p w^{1/p}.$$

Finally, letting

$$u \equiv R(h)^p w^{1/p} \qquad \text{and} \qquad v \equiv R(h)^{p'} w^{-1/p},$$

we have $u, v \in A_1$, $w = uv^{1-p}$, with

$$[u]_{A_1} \le c_n\,[w]_{A_p} \qquad \text{and} \qquad [v]_{A_1} \le c_n\,[w]_{A_p}^{\frac{1}{p-1}}. \qquad \square$$

3.3.2 The sharp extrapolation theorem

One of the main results in modern harmonic analysis is the extrapolation theorem of Rubio de Francia for A_p weights. This result is very useful because it reduces matters to studying one special exponent – typically $p = 2$. We refer the reader to the monograph [20] for a new proof and for a full account of the theory. On the other hand, in [26] there is a version of the extrapolation theorem with sharp constants which turns out to be very useful. The proof follows the classical method as exposed in [31] and it is based on the Rubio de Francia algorithm. Here we will give this proof.

Theorem 3.19. *Let T be any operator such that, for some exponent $\alpha > 0$,*

$$\|T\|_{L^2(w)} \le c\,[w]_{A_2}^{\alpha}, \qquad w \in A_2. \tag{3.31}$$

Then

$$\|T\|_{L^p(w)} \le c_p\,[w]_{A_p}^{\alpha}, \qquad p > 2, w \in A_p. \tag{3.32}$$

A corresponding result for $1 < p < 2$ can be found in [26], but since all the applications in these lecture notes deal with linear operators whose adjoints behave like the operator itself, we have the following corollary.

Corollary 3.20. *Let T be a linear operator satisfying (3.31). Suppose also that the adjoint operator T^* (with respect to the Lebesgue measure) also satisfies (3.31). Then*

$$\|T\|_{L^p(w)} \le c_p\,[w]_{A_p}^{\alpha \max\left\{1, \frac{1}{p-1}\right\}}, \qquad p > 1, w \in A_p.$$

The proof is simply a duality argument. All we have to do is to check the case $1 < p < 2$. Indeed, by standard theory,

$$\|T\|_{L^p(w)} = \|T^*\|_{L^{p'}(\sigma)},$$

where as usual $\sigma = w^{1-p'}$. Hence, if $1 < p < 2$ then $p' > 2$ and we can apply the theorem to T^* because it verifies (3.31), obtaining

$$\|T^*\|_{L^{p'}(\sigma)} \leq c\,[\sigma]^{\alpha}_{A_{p'}} = c\,[w]^{\frac{\alpha}{p-1}}_{A_p}.$$

Proof of Theorem 3.19. The proof that follows is taken from García-Cuerva and Rubio de Francia's book [31]. We only need to be absolutely precise with the exponents. Let

$$t = \frac{p-2}{p-1},$$

so that

$$\|T(f)\|^2_{L^p(w)} = \sup_h \int_{\mathbb{R}^n} |T(f)(x)|^2\, h(x)\, w(x)\, dx, \qquad (3.33)$$

where the supremum runs over all $0 \leq h \in L^{p'/t}(w)$ with $\|h\|_{L^{p'/t}(w)} = 1$.

We run the Rubio de Francia algorithm now as follows. Define the operator

$$S_w(h) = \left(\frac{M(h^{1/t}\,w)}{w}\right)^t, \qquad h \geq 0.$$

It is easy to see that, by Muckenhoupt's theorem, S_w is bounded on $L^{p'/t}(w)$ if $w \in A_p$. Furthermore, if we use Buckley's Theorem 3.11, we have

$$\|S_w\|_{L^{p'/t}(w)} \leq c_p\,[w]^t_{A_p}.$$

Define now

$$D(h) = \sum_{k=0}^{\infty} \frac{1}{2^k} \frac{S_w^k(h)}{\|S_w\|^k_{L^{p'/t}(w)}}.$$

Then we have:

(A) $h \leq D(h)$;
(B) $\|D(h)\|_{L^{p'/t}(w)} \leq 2\,\|h\|_{L^{p'/t}(w)}$;
(C) $S_w(D(h)) \leq 2\,\|S_w\|_{L^{p'/t}(w)}\,D(h)$, and hence

$$[D(h)\,w]_{A_2} \leq c\,[w]_{A_p}. \qquad (3.34)$$

Here (A) and (B) and the first part of (C) are immediate. It only remains to prove (3.34). First we claim that, for any $h \geq 0$,

$$[hw, S_w(h)w]_{A_2} = \sup_Q \left(\frac{1}{|Q|}\int_Q hw(x)\,dx\right)\left(\frac{1}{|Q|}\int_Q (S_w(h)w)^{-1}\,dx\right) \leq 2\,[w]^{1-t}_{A_p}.$$

Indeed, since for $x \in Q$ it verifies that

$$M(h^{1/t}w)(x) \geq \frac{1}{|Q|} \int_Q h^{1/t}w(x) \, dx,$$

we have that

$$\left(\frac{1}{|Q|} \int_Q hw(x) \, dx\right)\left(\frac{1}{|Q|} \int_Q (S_w(h)w)^{-1} \, dx\right)$$

$$= \left(\frac{1}{|Q|} \int_Q hw(x) \, dx\right)\left(\frac{1}{|Q|} \int_Q M(h^{1/t}w)^{-t}w^{-1/(p-1)} \, dx\right)$$

$$\leq \left(\frac{1}{|Q|} \int_Q h^{1/t}w(x) \, dx\right)^t \left(\frac{1}{|Q|} \int_Q w(x) \, dx\right)^{1-t}$$

$$\left(\frac{1}{|Q|} \int_Q h^{1/t}w(x) \, dx\right)^{-t} \left(\frac{1}{|Q|} \int_Q w(x)^{-1/(p-1)} \, dx\right)$$

$$\overset{?}{\leq} [w]_{A_p}^{1-t},$$

where we used Hölder's inequality and then that $(1-t)(1-p) = 1$. Hence,

$$[D(h) \, w]_{A_2} = \sup_Q \left(\frac{1}{|Q|} \int_Q D(h)w(x) \, dx\right)\left(\frac{1}{|Q|} \int_Q (D(h)w)^{-1} \, dx\right)$$

$$\leq 2 \, \|S_w\|_{L^{p'/t}(w)} \sup_Q \left(\frac{1}{|Q|} \int_Q D(h)w(x) \, dx\right)\left(\frac{1}{|Q|} \int_Q (S_w(D(h)) \, w)^{-1} \, dx\right)$$

$$= 2 \, \|S_w\|_{L^{p'/t}(w)} \, [D(h)w, S_w(D(h))w]_{A_2} \leq 2 \, \|S_w\|_{L^{p'/t}(w)} \, [w]_{A_p}^{1-t} \leq c \, [w]_{A_p}.$$

We are ready now to conclude the proof of the theorem using finally the extrapolation hypothesis (3.31). Indeed, fixing one of the h's from (3.33) we continue with

$$\int_{\mathbb{R}^n} |T(f)(x)|^2 \, h(x) \, w(x) \, dx \leq \int_{\mathbb{R}^n} |T(f)(x)|^2 \, D(h)(x) \, w(x) \, dx$$

$$\leq [D(h) \, w]_{A_2}^{2\alpha} \int_{\mathbb{R}^n} |f(x)|^2 \, D(h)(x) \, w(x) \, dx$$

$$\leq c \, [w]_{A_p}^{2\alpha} \int_{\mathbb{R}^n} |f(x)|^2 \, D(h)(x) \, w(x) \, dx$$

$$\leq c \, [w]_{A_p}^{2\alpha} \, \|f\|_{L^p(w)}^2 \, \|D(h)\|_{L^{(p/2)'}(w)} = c \, [w]_{A_p}^{2\alpha} \, \|f\|_{L^p(w)}^2 \, \|D(h)\|_{L^{p'/t}(w)}$$

$$\leq c \, [w]_{A_p}^{2\alpha} \, \|f\|_{L^p(w)}^2 \, \|h\|_{L^{p'/t}(w)} = c \, [w]_{A_p}^{2\alpha} \, \|f\|_{L^p(w)}^2.$$

This proves (3.32). □

3.4   Three more applications of Rubio de Francia's algorithm

We have seen in the previous section two applications of the Rubio de Francia algorithm. These applications were already known except for the use of the sharp exponents. In this section we show some other applications, perhaps more technical, but which play a crucial role in the works [52] and [54]. It should be mentioned that other applications can be found in [18], [19], [24] and [55]. Also we remit to the monograph [20] for more information.

3.4.1   Building A_1 weights from duality

The following lemma – a variation of the Rubio de Francia iteration scheme – is the key link for proving Conjecture 3.9 assuming that the A_1 conjecture holds.

Lemma 3.21. *Let $1 < p < \infty$ and let $w \in A_p$. Then there exists a nonnegative sublinear operator D bounded on $L^{p'}(w)$ such that, for any nonnegative $h \in L^{p'}(w)$,*

(a) $h \le D(h)$;

(b) $\|D(h)\|_{L^{p'}(w)} \le 2 \|h\|_{L^{p'}(w)}$;

(c) $D(h)\, w \in A_1$ *with* $[D(h)\, w]_{A_1} \le cp\, [w]_{A_p}$, *where the constant c is a dimensional constant.*

Proof. To define the algorithm, we consider the operator

$$S_w(f) = \frac{M(fw)}{w}$$

and observe that for any $1 < p < \infty$, by Muckenhoupt's theorem,

$$S_w \colon L^{p'}(w) \longrightarrow L^{p'}(w), \qquad w \in A_p.$$

However, we need the sharp version in both the constant and the A_p constant (3.17):

$$\|S_w\|_{L^{p'}(w)} \le cp\, [w^{1-p'}]_{A_{p'}}^{p-1} = cp\, [w]_{A_p}.$$

Define now, for any nonnegative $h \in L^{p'}(w)$,

$$D(h) = \sum_{k=0}^{\infty} \frac{1}{2^k} \frac{S_w^k(h)}{\|S_w\|_{L^{p'}(w)}^k}.$$

Hence, properties (a) and (b) are immediate, and for (c) simply observe that

$$S_w(D(h)) \le 2 \|S_w\|_{L^{p'}(w)} D(h) \le 2cp\, [w]_{A_p} D(h),$$

or, what is the same, $D(h)\, w \in A_1$ with

$$[D(h)\, w]_{A_1} \le 2cp\, [w]_{A_p}. \qquad \qquad \square$$

As an application of this lemma, we prove Conjecture 3.9 assuming that the A_1 conjecture (Conjecture 3.4) holds.

Proof of Conjecture 3.9. Let $w \in A_p$ and let $f \in C^\infty(\mathbb{R}^n)$ with compact support. For each $t > 0$, let

$$\Omega_t = \{x \in \mathbb{R}^n : |Tf(x)| > t\}.$$

This set is bounded, so $w(\Omega_t) < \infty$. By duality, there exists a nonnegative function $h \in L^{p'}(w)$ such that $\|h\|_{L^{p'}(w)} = 1$ and

$$w(\Omega_t)^{1/p} = \|\chi_{\Omega_t}\|_{L^p(w)} = \int_{\Omega_t} hw \, dx.$$

We now consider the operator D associated to this weight from Lemma 3.21. Thus the operator D satisfies

(a) $h \leq D(h)$;

(b) $\|Dh\|_{L^{p'}(w)} \leq 2 \|h\|_{L^{p'}(w)} = 2$;

(c) $[D(h)\,w]_{A_1} \leq cp\,[w]_{A_p}$.

Hence, assuming that the A_1 conjecture holds,

$$w(\Omega_t)^{1/p} \leq \int_{\Omega_t} D(h)\,w \, dx = (D(h)\,w)(\Omega_t)$$

$$\leq c\,[D(h)\,w]_{A_1} \int_{\mathbb{R}^n} \frac{|f|}{t} D(h)\,w \, dx$$

$$\leq \frac{c}{t} p\,[w]_{A_p} \left(\int_{\mathbb{R}^n} |f|^p w \, dx \right)^{1/p} \left(\int_{\mathbb{R}^n} D(h)^{p'} w \, dx \right)^{1/p'}$$

$$\leq \frac{cp}{t} [w]_{A_p} \left(\int_{\mathbb{R}^n} |f|^p w \, dx \right)^{1/p}.$$

This completes the proof. □

3.4.2 Improving inequalities with A_∞ weights

In harmonic analysis there are a number of important inequalities of the form

$$\int_{\mathbb{R}^n} |Tf(x)|^p \, w(x) \, dx \leq C \int_{\mathbb{R}^n} |Sf(x)|^p \, w(x) \, dx, \tag{3.35}$$

where T and S are operators. Typically, T is an operator with some degree of singularity (e.g., a singular integral operator), S is an operator which is, in principle, easier to handle (e.g., a maximal operator), and w is in some class of weights.

The standard technique for proving such results is the so-called good-λ inequality of Burkholder and Gundy. These inequalities compare the relative measure of the level sets of S and T: For every $\lambda > 0$ and $\epsilon > 0$ small,

$$w(\{y \in \mathbb{R}^n : |Tf(y)| > 2\lambda, |Sf(y)| \leq \lambda\epsilon\}) \tag{3.36}$$

$$\leq C\epsilon w(\{y \in \mathbb{R}^n : |Sf(y)| > \lambda\}).$$

Here, the weight w is usually assumed to be in the Muckenhoupt class $A_\infty = \cup_{p>1} A_p$. Given inequality (3.36), it is easy to prove the strong-type inequality (3.35) for any p, $0 < p < \infty$, as well as the corresponding weak-type inequality

$$\|Tf\|_{L^{p,\infty}(w)} \leq C \|Sf\|_{L^{p,\infty}(w)}. \tag{3.37}$$

In these notes the special case of

$$\|Tf\|_{L^p(w)} \leq c \|Mf\|_{L^p(w)}, \tag{3.38}$$

where T is a Calderón–Zygmund operator and M is the maximal function, will play a central role. This theorem was proved by Coifman–Fefferman in the celebrated paper [14]. Here the weight w also satisfies the A_∞ condition, but the problem is that the behavior of the constant is too rough. We need a more precise result for very specific weights.

Lemma 3.22 (The tricky lemma). *Let w be any weight and let $1 \leq p, r < \infty$. Then there is a constant $c = c(n, T)$ such that*

$$\|Tf\|_{L^p(M_r w)^{1-p}} \leq cp \|Mf\|_{L^p(M_r w)^{1-p}}.$$

This is the main improvement in [54] of [52], where we had obtained logarithmic growth on p. It is an important step towards the proof of the linear growth theorem (Theorem 3.5).

The above-mentioned good λ of Coifman–Fefferman is not sharp, since instead of cp it gives $C(p) \approx 2^p$, because

$$[(M_r w)^{1-p})]_{A_p} \approx (r')^{p-1}.$$

There is another proof by R. Bagby and D. Kurtz using rearrangements given in the middle of the 80's, which gives better estimates on p but not in terms of the weight constant.

The proof of this lemma is tricky and combines another variation of the Rubio de Francia algorithm together with a sharp L^1 version of (3.38):

$$\|Tf\|_{L^1(w)} \leq c [w]_{A_q} \|Mf\|_{L^1(w)}, \qquad w \in A_q, \ 1 \leq q < \infty. \tag{3.39}$$

The original proof given in [54] of this estimate was based on an idea of Fefferman–Pipher from [28], which combines a sharp version of the good-λ inequality of Buckley together with a sharp reverse Hölder property of the weights

(Lemma 3.28). The result of Buckley establishes a very interesting exponential improvement of the good-λ estimate of the above-mentioned Coifman–Fefferman estimate, as can be found in [8]:

$$|\{x \in \mathbb{R}^n : T^*(f) > 2\lambda, \, Mf < \gamma\lambda\}| \leq c_1 e^{-c_2/\gamma} |\{T^*(f) > \lambda\}|, \quad \lambda, \gamma > 0, \quad (3.40)$$

where T^* is the maximal singular integral operator. This approach is interesting on its own, but we will present in these lecture notes a more efficient approach based on the following estimate: Let $0 < p < \infty$, $0 < \delta < 1$, and let $w \in A_q$, $1 \leq q < \infty$. Then

$$\|f\|_{L^p(w)} \leq c p \, [w]_{A_q} \|M_\delta^\#(f)\|_{L^p(w)} \quad (3.41)$$

for any function f such that $|\{x : |f(x)| > t\}| < \infty$. Here,

$$M_\delta^\# f(x) = M^\#(|f|^\delta)(x)^{1/\delta}$$

and $M^\#$ is the usual sharp maximal function of Fefferman–Stein:

$$M^\#(f)(x) = \sup_{Q \ni x} \frac{1}{|Q|} \int_Q |f(y) - f_Q| \, dy,$$

with $f_Q = \frac{1}{|Q|} \int_Q f(y) \, dy$. We present this theory in Section 3.11.

To prove (3.39), we combine (3.41) with the following pointwise estimate [2]:

Lemma 3.23. *Let T be any Calderón–Zygmund singular integral operator and let $0 < \delta < 1$. Then there is a constant c such that*

$$M_\delta^\#(T(f))(x) \leq c \, Mf(x).$$

In fact, we have:

Corollary 3.24. *Let $0 < p < \infty$ and let $w \in A_q$. Then*

$$\|Tf\|_{L^p(w)} \leq c \, [w]_{A_q} \|Mf\|_{L^p(w)}$$

for any f such that $|\{x : |Tf(x)| > t\}| < \infty$.

Hence the heart of the matter is the estimate (3.41). It will be proved in Section 3.11 (see Corollary 3.43) by a completely different path, using instead properties of rearrangement of functions and corresponding local sharp maximal operator.

We now finish this section by proving the "tricky" Lemma 3.22. The proof is based on the following lemma, which is another variation of the Rubio de Francia algorithm.

Lemma 3.25. *Let $1 < s < \infty$ and let w be a weight. Then there exists a nonnegative sublinear operator R satisfying the following properties:*

(a) $h \leq R(h)$;

(b) $\|R(h)\|_{L^s(w)} \leq 2 \|h\|_{L^s(w)}$;

(c) $R(h)w^{1/s} \in A_1$ *with*

$$[R(h)w^{1/s}]_{A_1} \leq cs'.$$

Proof. We consider the operator

$$S(f) = \frac{M(f\,w^{1/s})}{w^{1/s}}.$$

Since $\|M\|_{L^s} \sim s'$, we have

$$\|S(f)\|_{L^s(w)} \leq cs'\|f\|_{L^s(w)}.$$

Now, define the Rubio de Francia operator R by

$$R(h) = \sum_{k=0}^{\infty} \frac{1}{2^k} \frac{S^k(h)}{\left(\|S\|_{L^s(w)}\right)^k}.$$

It is very simple to check that R satisfies the required properties. $\square$

Proof of Lemma 3.22. We are now ready to give the proof of the "tricky" Lemma, namely to prove that

$$\left\|\frac{Tf}{M_r w}\right\|_{L^p(M_r w)} \leq cp \left\|\frac{Mf}{M_r w}\right\|_{L^p(M_r w)}.$$

By duality we have

$$\left\|\frac{Tf}{M_r w}\right\|_{L^p(M_r w)} = \left|\int_{\mathbb{R}^n} Tf\,h\,dx\right| \leq \int_{\mathbb{R}^n} |Tf|\,h\,dx$$

for some h with $\|h\|_{L^{p'}(M_r w)} = 1$. By Lemma 3.25 with $s = p'$ and $v = M_r w$, there exists an operator R such that

(A) $h \leq R(h)$;

(B) $\|R(h)\|_{L^{p'}(M_r w)} \leq 2 \|h\|_{L^{p'}(M_r w)}$;

(C) $[R(h)(M_r w)^{1/p'}]_{A_1} \leq cp$.

We want to make use of property (C) combined with the following two facts: First, if $w_1, w_2 \in A_1$ and $w = w_1 w_2^{1-p} \in A_p$, then, by (3.29),

$$[w]_{A_p} \leq [w_1]_{A_1}[w_2]_{A_1}^{p-1}.$$

Second, if $r > 1$ then $(Mf)^{\frac{1}{r}} \in A_1$ by the Coifman–Rochberg theorem. Furthermore, we need to be more precise (3.42):

$$[(Mf)^{\frac{1}{r}}]_{A_1} \le c_n r'.$$

Hence, combining, we obtain

$$[R(h)]_{A_3} = \left[R(h)(M_r w)^{1/p'} \left((M_r w)^{1/2p'} \right)^{-2} \right]_{A_3}$$

$$\le \left[R(h)(M_r w)^{1/p'} \right]_{A_1} \left[(M_r w)^{1/2p'} \right]_{A_1}^2$$

$$\le cp.$$

Therefore, by Corollary 3.24 and by properties (A) and (B),

$$\int_{\mathbb{R}^n} |Tf|\, h\, dx \le \int_{\mathbb{R}^n} |Tf|\, R(h)\, dx \le c\, \|R(h)\|_{A_3} \int_{\mathbb{R}^n} M(f) R(h)\, dx$$

$$\le cp \left\| \frac{Mf}{M_r w} \right\|_{L^p(M_r w)} \|h\|_{L^{p'}(M_r w)}. \qquad \square$$

3.5   The sharp reverse Hölder property of A_1 weights

We already encountered weights of the form $M_r w$, $1 < r < \infty$. As we are going to see, they play an important role in the theory. Of course, it is well known that these weights satisfy the A_1 condition by the theorem of Coifman–Rochberg [15]. We will be using the following quantitative version of it: Let μ be a positive Borel measure and let $1 < r < \infty$. Then

$$(M\mu)^{\frac{1}{r}} \in A_1$$

and furthermore

$$[(M\mu)^{\frac{1}{r}}]_{A_1} \le c_n r'. \tag{3.42}$$

In fact, they proved that any A_1 weight can be essentially written in this way.

Recall that these weights satisfy a special important property, namely that, if $w \in A_1$, then there is a constant $r > 1$ such that

$$\left(\frac{1}{|Q|} \int_Q w^r \right)^{1/r} \le \frac{c}{|Q|} \int_Q w.$$

However, there is a bad dependence on the constant $c = c(r, [w]_{A_1})$. To prove our results, we need a more precise estimate.

Lemma 3.26. *Let* $w \in A_1$ *and let* $r_w = 1 + \frac{1}{2^{n+1}[w]_{A_1}}$. *Then, for each* Q,

$$\left(\frac{1}{|Q|} \int_Q w^{r_w} \right)^{1/r_w} \le \frac{2}{|Q|} \int_Q w,$$

i.e., $M_{r_w} w(x) \le 2 [w]_{A_1} w(x)$.

(Recall that $M_r w = M(w^r)^{1/r}$.)

Proof. We start with the "layer cake formula"

$$\int_X \varphi(f) \, d\nu = \int_0^\infty \varphi'(t) \, \nu(\{x \in X : f(x) > t\}) \, dt.$$

We fix a cube Q and denote by M_Q^d the dyadic maximal operator restricted to the cube Q. Also we denote $w_Q = \frac{1}{|Q|} \int_Q w$. Hence,

$$\frac{1}{|Q|} \int_Q w^{1+\delta} \, dx \le \frac{1}{|Q|} \int_Q M_Q^d(w)(x)^\delta \, w \, dx$$

$$= \frac{\delta}{|Q|} \int_0^\infty t^{\delta-1} \, w(\{x \in Q : M_Q^d(w)(x) > t\}) \, dt$$

$$\le \frac{\delta}{|Q|} \int_0^{w_Q} t^{\delta-1} \, w(\{x \in Q : M_Q^d(w)(x) > t\}) \, dt$$

$$+ \frac{\delta}{|Q|} \int_{w_Q}^\infty t^{\delta-1} \, w(\{x \in Q : M_Q^d(w)(x) > t\}) \, dt$$

$$\le (w_Q)^{\delta+1} + \frac{\delta}{|Q|} \int_{w_Q}^\infty t^{\delta-1} \, w(\{x \in Q : M_Q^d(w)(x) > t\}) \, dt.$$

We now use a sort of "reverse" weak type $(1,1)$ estimate for the maximal function: If $t > w_Q$, then

$$w(\{x \in Q : M_Q^d w(x) > t\}) \le 2^n t \, |\{x \in Q : M_Q^d w(x) > t\}|,$$

that can be found, for instance, in [32]. Hence,

$$\frac{1}{|Q|} \int_Q (M_Q^d w)^\delta w \, dx \le (w_Q)^{\delta+1} + \frac{2^n \delta}{\delta+1} \frac{1}{|Q|} \int_Q (M_Q^d w)^{\delta+1} \, dx$$

$$\le (w_Q)^{\delta+1} + \frac{2^n \delta [w]_{A_1}}{\delta+1} \frac{1}{|Q|} \int_Q (M_Q^d w)^\delta w \, dx.$$

Setting here $\delta = \frac{1}{2^{n+1}[w]_{A_1}}$, we obtain

$$\frac{1}{|Q|} \int_Q (M_Q^d w)^\delta w \, dx \le 2(w_Q)^{\delta+1}. \qquad \square$$

3.6 Main lemma and proof of the linear growth theorem

In this section we combine all the previous information to finish the proof of the linear growth theorem (Theorem 3.5). We need a lemma which immediately gives the proof.

Lemma 3.27. *Let T be any Calderón–Zygmund singular integral operator and let w be any weight. Also let $1 < p < \infty$ and $1 < r < 2$. Then there is a $c = c_n$ such that*

$$\|Tf\|_{L^p(w)} \le cp' \left(\frac{1}{r-1}\right)^{1-1/pr} \|f\|_{L^p(M_r w)}.$$

In applications we will often use the following consequence:

$$\|Tf\|_{L^p(w)} \le cp' \, (r')^{1/p'} \|f\|_{L^p(M_r w)}$$

since $t^{1/t} \le 2$ for $t \ge 1$.

We are now ready to finish the proof of the linear growth theorem.

Proof of Theorem 3.5. Indeed, apply the lemma to $w \in A_1$ with sharp reverse Hölder exponent $r = r_w = 1 + \frac{1}{2^{n+1}[w]_{A_1}}$, obtaining

$$\|T\|_{L^p(w)} \le cp' \, [w]_{A_1}. \qquad \square$$

Proof of the lemma. We consider the equivalent dual estimate:

$$\|T^* f\|_{L^{p'}(M_r w)^{1-p'}} \le cp' \left(\frac{1}{r-1}\right)^{1-1/pr} \|f\|_{L^{p'}(w^{1-p'})}.$$

Then use the "tricky" Lemma 3.22, since T^* is also a Calderón–Zygmund operator:

$$\left\| \frac{T^* f}{M_r w} \right\|_{L^{p'}(M_r w)} \le p'c \left\| \frac{Mf}{M_r w} \right\|_{L^{p'}(M_r w)}.$$

Next we note that, by Hölder's inequality with exponent pr,

$$\frac{1}{|Q|} \int_Q f w^{-1/p} w^{1/p} \le \left(\frac{1}{|Q|} \int_Q w^r \right)^{1/pr} \left(\frac{1}{|Q|} \int_Q (f w^{-1/p})^{(pr)'} \right)^{1/(pr)'}$$

and hence

$$(Mf)^{p'} \le (M_r w)^{p'-1} M\left((f w^{-1/p})^{(pr)'} \right)^{p'/(pr)'}.$$

From this, and by the classical unweighted maximal theorem with sharp constant,

$$\left\| \frac{Mf}{M_r w} \right\|_{L^{p'}(M_r w)} \le c \left(\frac{p'}{p' - (pr)'} \right)^{1/(pr)'} \left\| \frac{f}{w} \right\|_{L^{p'}(w)}$$

$$= c \left(\frac{rp-1}{r-1} \right)^{1-1/pr} \left\| \frac{f}{w} \right\|_{L^{p'}(w)} \le cp \left(\frac{1}{r-1} \right)^{1-1/pr} \left\| \frac{f}{w} \right\|_{L^{p'}(w)}. \qquad \square$$

3.7   Proof of the logarithmic growth theorem

Proof of Theorem 3.6. The proof is based on ideas from [64]. Applying the Calderón–Zygmund decomposition to f at level λ, we get a family of pairwise disjoint cubes $\{Q_j\}$ such that

$$\lambda < \frac{1}{|Q_j|} \int_{Q_j} |f| \le 2^n \lambda.$$

Let $\Omega = \cup_j Q_j$ and $\widetilde{\Omega} = \cup_j 2Q_j$. The "good part" is defined by

$$g = \sum_j f_{Q_j} \chi_{Q_j}(x) + f(x)\chi_{\Omega^c}(x)$$

and the "bad part" by

$$b = \sum_j b_j, \qquad \text{where} \qquad b_j(x) = (f(x) - f_{Q_j})\chi_{Q_j}(x).$$

Then $f = g + b$.

However, it turns out that b is "excellent" and g is really "ugly". The b part is so good that we have the full Muckenhoupt–Wheeden conjecture:

$$w(\{x \in (\widetilde{\Omega})^c : |Tb(x)| > \lambda\}) \le \frac{c}{\lambda} \int_{\mathbb{R}^n} |f|\, Mw\, dx$$

by a well-known argument using the cancelation of the b_j, that we omit. Also the term $w(\widetilde{\Omega})$ is the level set of the maximal function and the Fefferman–Stein estimate applies (again we have the full Muckenhoupt conjecture).

Combining we have

$$w(\{x \in \mathbb{R}^n : |Tf(x)| > \lambda\}) \le w(\widetilde{\Omega}) + w(\{x \in (\widetilde{\Omega})^c : |Tb(x)| > \lambda/2\})$$

$$+ w(\{x \in (\widetilde{\Omega})^c : |Tg(x)| > \lambda/2\}),$$

and the first two terms are already controlled:

$$w(\widetilde{\Omega}) + w(\{x \in (\widetilde{\Omega})^c : |Tb(x)| > \lambda/2\}) \le \frac{c}{\lambda} \int_{\mathbb{R}^n} |f|\, Mw\, dx \le \frac{c[w]_{A_1}}{\lambda} \int_{\mathbb{R}^n} |f|\, w\, dx.$$

Now, by Chebyshev and the lemma, for any $p > 1$ we have

$$w(\{x \in (\widetilde{\Omega})^c : |Tg(x)| > \lambda/2\})$$

$$\le c(p')^p \Big(\frac{1}{r-1}\Big)^{p-\frac{1}{r}} \frac{1}{\lambda^p} \int_{\mathbb{R}^n} |g|^p M_r(w\chi_{(\widetilde{\Omega})^c})\, dx$$

$$\le c(p')^p \Big(\frac{1}{r-1}\Big)^{p-\frac{1}{r}} \frac{1}{\lambda} \int_{\mathbb{R}^n} |g|\, M_r(w\chi_{(\widetilde{\Omega})^c})\, dx.$$

By more or less standard arguments,

$$\int_{\mathbb{R}^n} |g| \, M_r(w\chi_{(\widetilde{\Omega})^c}) \, dx \leq c \int_{\mathbb{R}^n} |f| \, M_r w \, dx.$$

Combining this estimate with the previous one, and then taking the sharp reverse Hölder exponent $r = 1 + \frac{1}{2^{n+1}[w]_{A_1}}$, by the reverse Hölder inequality lemma we get

$$w(\{x \in (\widetilde{\Omega})^c \, : \, |Tg(x)| > \lambda/2\}) \leq \frac{c(p'[w]_{A_1})^p}{\lambda} \int_{\mathbb{R}^n} |f| \, w \, dx.$$

Setting here

$$p = 1 + \frac{1}{\log(1 + [w]_{A_1})}$$

gives

$$w(\{x \in (\widetilde{\Omega})^c \, : \, |Tg(x)| > \lambda/2\}) \leq \frac{c[w]_{A_1}(1 + \log[w]_{A_1})}{\lambda} \int_{\mathbb{R}^n} |f| \, w \, dx.$$

This estimate combined with the previous one completes the proof. $\qquad\square$

3.8 Properties of A_p weights

We have seen how important is the sharp reverse Hölder exponent for A_1 weights (Lemma 3.26) for the proof of Theorem 3.5 and hence for that of Theorem 3.6. A natural question is then to find a similar result for the A_p class of weights. The question is interesting on its own but it turns out that it is very useful as well as can be seen in the proof of the quadratic estimates for commutators given in Section 3.10.

Recall that if $w \in A_p$ then there are constants $r > 1$ and $c \geq 1$ such that, for any cube Q,

$$\left(\frac{1}{|Q|} \int_Q w^r \, dx\right)^{\frac{1}{r}} \leq \frac{c}{|Q|} \int_Q w. \tag{3.43}$$

In the standard proofs, both constants c, r depend upon the A_p constant of the weight. We prove here a more precise version of (3.43).

Lemma 3.28. *Let $w \in A_p$, $1 < p < \infty$, and let $r_w = 1 + \frac{1}{2^{2p+n+1}[w]_{A_p}}$. Then, for any Q,*

$$\left(\frac{1}{|Q|} \int_Q w^{r_w} \, dx\right)^{\frac{1}{r_w}} \leq \frac{2}{|Q|} \int_Q w. \tag{3.44}$$

Remark 3.29. We remark that this result has been considerably improved in [38], as stated below in Theorem 3.31, although we skip the proof, which is different from the one presented here after the next corollary.

The classical proofs of this property for any A_p weights produce nonlinear growth constants. We also remark that this result was stated and used in [8] with no proof. The author mentioned instead the work by Coifman–Fefferman [14], where no explicit statement can be found. Since this lemma plays an important role (especially the case $p = 2$), we supply below a proof.

As a corollary, we deduce the following useful result.

Corollary 3.30. *Let $1 < p < \infty$ and let $w \in A_p$. Denote*

$$p_w = \frac{p}{1 + \dfrac{p-1}{1 + \dfrac{1}{2^{2p'+n+1}[w]_{A_p}^{p'-1}}}}.$$

Then $w \in A_{p/p_w}$ and furthermore

$$[w]_{A_{p/p_w}} \leq 2^{p-1}[w]_{A_p},$$

or, equivalently, $w \in A_{p-\epsilon}$ where $\epsilon = \dfrac{p-1}{1 + 2^{2p'+n+1}[w]_{A_p}^{p'-1}} \approx \dfrac{p-1}{[w]_{A_p}^{p'-1}}.$

The proof of the corollary is as follows. Since $w \in A_p$, $\sigma \in A_{p'}$, and hence, by the lemma,

$$r_\sigma = 1 + \frac{1}{2^{2p'+n+1}[\sigma]_{A_{p'}}} = 1 + \frac{1}{2^{2p'+n+1}[w]_{A_p}^{p'-1}}$$

with

$$\left(\frac{1}{|Q|}\int_Q \sigma^{r_\sigma}\,dx\right)^{\frac{1}{r_\sigma}} \leq \frac{2}{|Q|}\int_Q \sigma$$

and therefore

$$\left(\frac{1}{|Q|}\int_Q w(x)\,dx\right)\left(\frac{1}{|Q|}\int_Q \sigma(x)^{r_\sigma}\,dx\right)^{\frac{p-1}{r_\sigma}}$$

$$\leq 2^{p-1}\left(\frac{1}{|Q|}\int_Q w(x)\,dx\right)\left(\frac{2}{|Q|}\int_Q w^{1-p'}\,dx\right)^{p-1},$$

namely

$$[w]_{A_{p/p_w}} \leq 2^{p-1}[w]_{A_p}$$

since

$$\frac{p-1}{r_\sigma} = \frac{p}{p_w} - 1.$$

This theorem plays an important role in deriving a sharp version of the Coifman–Fefferman estimate as stated in Corollary 3.44, which plays a central role in (3.5).

Proof. Let $w_Q = \frac{1}{|Q|} \int_Q w$. Then

$$\frac{1}{|Q|} \int_Q w(x)^\delta \, w(x) \, dx = \frac{\delta}{|Q|} \int_0^\infty t^\delta w(\{x \in Q \, : \, w(x) > t\}) \frac{dt}{t}$$

$$= \frac{\delta}{|Q|} \int_0^{w_Q} t^\delta w(\{x \in Q \, : \, w(x) > t\}) \frac{dt}{t}$$

$$+ \frac{\delta}{|Q|} \int_{w_Q}^\infty t^\delta w(\{x \in Q \, : \, w(x) > t\}) \frac{dt}{t} = I + II.$$

Observe that $I \le (w_Q)^{\delta+1}$. For II we make first the following observation: For any Q, we let

$$E_Q = \left\{ x \in Q \, : \, w(x) \le \frac{1}{2^{p-1}[w]_{A_p}} w_Q \right\}.$$

Then we claim that

$$|E_Q| \le \frac{1}{2}|Q|. \tag{3.45}$$

Indeed, by Hölder's inequality we have, for any $f \ge 0$,

$$\left(\frac{1}{|Q|} \int_Q f(y) \, dy \right)^p w(Q) \le [w]_{A_p} \int_Q f(y)^p \, w(y) \, dy$$

and hence, if $E \subset Q$,

$$\left(\frac{|E|}{|Q|} \right)^p \le [w]_{A_p} \frac{w(E)}{w(Q)}$$

and, in particular,

$$\left(\frac{|E_Q|}{|Q|} \right)^p \le [w]_{A_p} \frac{w(E_Q)}{w(Q)} \le [w]_{A_p} \frac{w_Q}{w(Q)} |E_Q| \frac{1}{2^{p-1}[w]_{A_p}} = \frac{1}{2^{p-1}} \frac{|E_Q|}{|Q|},$$

from which the claim follows.

The second claim is the following: For every $\lambda > w_Q$,

$$w(\{x \in Q \, : \, w(x) > \lambda\}) \le 2^{n+1}\lambda \left| \left\{ x \in Q \, : \, w(x) > \frac{\lambda}{2^{p-1}[w]_{A_p}} w_Q \right\} \right|. \tag{3.46}$$

Consider the CZ decomposition of w at level λ. We find a family of disjoint cubes $\{Q_i\}$ contained in Q satisfying

$$\lambda < w_{Q_i} \le 2^n \lambda$$

for each i. Indeed, except for a null set we have

$$\{x \in Q \, : \, w(x) > \lambda\} \subset \{x \in Q \, : \, M_Q^d \, w(x) > \lambda\} = \cup_i Q_i,$$

where M_Q^d is the dyadic maximal operator restricted to a cube Q. Hence this, together with (3.45), yields

$$w(\{x \in Q : w(x) > \lambda\}) \le \sum_i w(Q_i) \le 2^n \lambda \sum_i |Q_i|$$

$$\le 2^{n+1} \lambda \sum_i \left| \left\{ x \in Q_i : w(x) > \frac{1}{2^{p-1}[w]_{A_p}} w_{Q_i} \right\} \right|$$

$$\le 2^{n+1} \lambda \left| \left\{ x \in Q : w(x) > \frac{1}{2^{p-1}[w]_{A_p}} \lambda \right\} \right|,$$

since $w_{Q_i} > \lambda$. This proves claim (3.46).

$$II = \frac{\delta}{|Q|} \int_{w_Q}^{\infty} t^\delta w(\{x \in Q : w(x) > \lambda\}) \frac{d\lambda}{\lambda}$$

$$\le \frac{2^{n+1}\delta}{|Q|} \int_{w_Q}^{\infty} \lambda^{\delta+1} \left| \left\{ x \in Q : w(x) > \frac{1}{2^{p-1}[w]_{A_p}} \lambda \right\} \right| \frac{d\lambda}{\lambda}$$

$$\le \left(2^{p-1}[w]_{A_p} \right)^{1+\delta} 2^{n+1} \delta \frac{1}{|Q|} \int_{\frac{w_Q}{2^{p-1}[w]_{A_p}}}^{\infty} \lambda^{\delta+1} |\{x \in Q : w(x) > \lambda\}| \frac{d\lambda}{\lambda}$$

$$\le \left(2^{p-1}[w]_{A_p} \right)^{1+\delta} 2^{n+1} \frac{\delta}{1+\delta} \frac{1}{|Q|} \int_Q w^{1+\delta} \, dx.$$

Setting here $\delta = \dfrac{1}{2^{2p+n+1}[w]_{A_p}}$, we obtain, using that $t^{1/t} \le e$ for $t \ge 1$,

$$\frac{1}{|Q|} \int_Q w^{\delta+1} \, dx \le 2(w_Q)^{\delta+1},$$

which proves (3.44).   □

3.9   Improvements in terms of mixed A_1-A_∞ constants

Since the A_p classes are increasing with respect to p, we can define the A_∞ class in a natural way by

$$A_\infty = \bigcup_{p>1} A_p.$$

For any weight in this larger class, the A_∞ constant can be defined as follows:

$$\|w\|_{A_\infty} = \sup_Q \left(\frac{1}{|Q|} \int_Q w \right) \exp \left(\frac{1}{|Q|} \int_Q \log w^{-1} \right).$$

This constant was introduced by Hruščev [35] (see also [31]) and has been the standard A_∞ constant until very recently, when a "new" A_∞ constant has been found to be better suited. This new constant is defined as

$$[w]_{A_\infty} = \sup_Q \frac{1}{w(Q)} \int_Q M(w\chi_Q). \tag{3.47}$$

In fact, this constant was introduced by M. Wilson long time ago (see [78, 79, 80]) with a different notation. This constant is more relevant since there are examples of weights $w \in A_\infty$ such that $[w]_{A_\infty}$ is much smaller than $\|w\|_{A_\infty}$. Indeed, it can be shown that

$$c_n [w]_{A_\infty} \leq \|w\|_{A_\infty} \leq [w]_{A_p}, \qquad 1 < p < \infty, \tag{3.48}$$

where c_n is a constant depending only on the dimension. The first inequality is the nontrivial part and can be found in [38], where a more interesting fact is also shown, namely that this inequality can be strict. More precisely, the authors exhibit a family of weights $\{w_t\}$ such that $[w_t]_{A_\infty} \leq 4\log(t)$ and $\|w_t\|_{A_\infty} \sim t/\log(t)$ for $t \gg 1$. It should be mentioned that this constant has also been used by Lerner [51, Section 5.5] (see also [42]) where the term "A_∞ constant" was coined for the first time.

In this section, we state a new result obtained by the author with T. Hytönen in [38]. This sharper version of the reverse Hölder inequality plays a central role in the proofs of all these new results.

Theorem 3.31 (A new sharp reverse Hölder inequality). *Define*

$$r_w = 1 + \frac{1}{\tau_n [w]_{A_\infty}},$$

where τ_n is a dimensional constant that we may take to be $\tau_n = 2^{11+n}$. Note that $r'_w \approx [w]_{A_\infty}$.

(a) *If $w \in A_\infty$, then*

$$\left(\frac{1}{|Q|} \int_Q w^{r_w} \right)^{1/r_w} \leq \frac{2}{|Q|} \int_Q w.$$

(b) *Furthermore, the result is optimal up to a dimensional factor: If a weight w satisfies the reverse Hölder inequality, i.e., there exists a constant K such that*

$$\left(\frac{1}{|Q|} \int_Q w^r \right)^{1/r} \leq \frac{K}{|Q|} \int_Q w,$$

then there exists a dimensional constant $c = c_n$ such that $[w]_{A_\infty} \leq c_n K r'$.

Remark 3.32. Results analogous to the left-hand side of inequality (3.48) and Theorem 3.31 have been independently obtained by O. Beznosova and A. Reznikov in [7]. Their formulation is slightly different and involves yet another weight constant closely related to $[w]_{A_\infty}$.

We finish this section by stating an improvement of Theorem 3.5 obtained in [38] using this new sharp reverse Hölder property. The new idea is to derive results mixing both constants A_1 and A_∞ in the final result.

Theorem 3.33. *Let T be a Calderón–Zygmund operator and let $1 < p < \infty$. Then*

$$\|T\|_{L^p(w)} \leq c\, pp'\, [w]_{A_1}^{1/p} [w]_{A_\infty}^{1/p'}, \qquad w \in A_1, \quad 1 < p < \infty,$$

and

$$\|T\|_{L^{1,\infty}(w)} \leq c\, [w]_{A_1} \log(e + [w]_{A_\infty}) \|f\|_{L^1(w)}.$$

In view of [58], this seems to be the best possible result.

3.10 Quadratic estimates for commutators

In this section we consider commutators of singular integrals with bounded mean oscillation (BMO) functions. When T is a singular integral operator, these operators were considered by Coifman, Rochberg and Weiss in [16]. Formally these operators are defined by

$$[b, T]f(x) = b(x)T(f)(x) - T(bf)(x) = \int_{\mathbb{R}^n} (b(x) - b(y))K(x, y)f(y)\, dy,$$

where K is a kernel satisfying the standard Calderón–Zygmund estimates. Although the original interest in the study of such operators was related to generalizations of the classical factorization theorem for Hardy spaces, many other applications have been found.

The main result from [16] states that $[b, T]$ is a bounded operator on $L^p(\mathbb{R}^n)$, $1 < p < \infty$, when b is a BMO function and T is a singular integral operator. In fact, the BMO condition of b is also a necessary condition for the L^p-boundedness of the commutator when T is the Hilbert transform. We may think that these operators behave as Calderón–Zygmund operators, although there are some differences. For instance, simple examples show that in general $[b, T]$ fails to be of weak type $(1, 1)$ when $b \in BMO$. This was observed by the author in [65], where it is also shown that there is an appropriate weak-$L(\log L)$ type estimate replacement. To stress this point of view it is also shown by the author [66] that the right operator controlling $[b, T]$ is $M^2 = M \circ M$, instead of the Hardy–Littlewood maximal function M. We pursue in this way by showing that commutators have an extra "bad" behavior from the point of view of A_p weights when trying to derive theorems such as Theorems 3.5 or 3.6.

3.10.1 A preliminary result: a sharp connection between the John–Nirenberg theorem and the A_2 class

In this section we outline the main results from [13]. We want to stress how important are the reverse Hölder property of the A_2 weights in conjunction with the following sharp version of the classical and well-known John–Nirenberg theorem.

For a locally integrable $b\colon \mathbb{R}^n \to \mathbb{R}$ we define

$$\|b\|_{BMO} = \sup_Q \frac{1}{|Q|} \int_Q |b(y) - b_Q|\, dy < \infty,$$

where the supremum is taken over all cubes $Q \in \mathbb{R}^n$ with sides parallel to the axes, and

$$b_Q = \frac{1}{|Q|} \int_Q b(y)\, dy.$$

The main relevance of BMO is due to the fact that it has an exponential self-improving property, namely the celebrated John–Nirenberg theorem. We need a very precise version of it, as follows:

Theorem 3.34 (Sharp John–Nirenberg inequality). *There are dimensional constants $0 \leq \alpha_n < 1 < \beta_n$ such that*

$$\sup_Q \frac{1}{|Q|} \int_Q \exp\left(\frac{\alpha_n}{\|b\|_{BMO}} |b(y) - b_Q|\right)\, dy \leq \beta_n. \tag{3.49}$$

In fact we can take $\alpha_n = \frac{1}{2^{n+2}}$.

For the proof of this we refer to the lecture notes by J. L. Journé [41, pp. 31–32]. The result there is not so explicit but it follows from the proof, which is very interesting and different from the usual ones that can be found in many references.

We derive from Theorem 3.34 the following relationship between BMO and the A_2 class of weights (Lemma 3.35), that will be used in the proof of the main theorem of this section. Indeed, it is well known that if $w \in A_2$ then $b = \log w \in BMO$. A partial converse also holds, namely if $b \in BMO$ then there is an $s_0 > 0$ such that $w = e^{sb} \in A_p$ if $|s| \leq s_0$. We have now a more precise version of this converse.

Lemma 3.35. *Let $b \in BMO$ and let $\alpha_n < 1 < \beta_n$ be the dimensional constants from (3.49). If $s \in \mathbb{R}$ and*

$$|s| \leq \frac{\alpha_n}{\|b\|_{BMO}},$$

then $e^{sb} \in A_2$ and $[e^{sb}]_{A_2} \leq \beta_n^2$.

Proof. By Theorem 3.34, if $|s| \leq \frac{\alpha_n}{\|b\|_{BMO}}$ and Q is fixed,

$$\frac{1}{|Q|} \int_Q \exp(|s||b(y) - b_Q|)\, dy \leq \frac{1}{|Q|} \int_Q \exp\left(\frac{\alpha_n}{\|b\|_{BMO}} |b(y) - b_Q|\right)\, dy \leq \beta_n$$

and then

$$\frac{1}{|Q|} \int_Q \exp(s(b(y) - b_Q))\, dy \leq \beta_n$$

and

$$\frac{1}{|Q|} \int_Q \exp(-s(b(y) - b_Q)) \, dy \leq \beta_n.$$

If we multiply the inequalities, the b_Q parts cancel out:

$$\left(\frac{1}{|Q|} \int_Q \exp(s(b(y) - b_Q)) \, dy \right) \left(\frac{1}{|Q|} \int_Q \exp(s(b_Q - b(y))) \, dy \right)$$

$$= \left(\frac{1}{|Q|} \int_Q \exp(sb(y)) \, dy \right) \left(\frac{1}{|Q|} \int_Q \exp(-sb(y)) \, dy \right) \leq \beta_n^2,$$

namely $e^{sb} \in A_2$ with $[e^{sb}]_{A_2} \leq \beta_n^2$.   □

3.10.2   Results in the A_p context

In this section we use the extrapolation theorem (Corollary 3.20) to get the optimal estimates for commutators. We remark that the next results are stated for very general operators. These results can be found in [13], as well as some generalizations.

Theorem 3.36. *Let T be a linear operator such that*

$$\|T\|_{L^2(w)} \leq c \, [w]_{A_2}, \quad w \in A_2. \tag{3.50}$$

Then there is a constant c independent of w and b such that

$$\|[b, T]\|_{L^2(w)} \leq c \, [w]_{A_2}^2 \, \|b\|_{BMO}. \tag{3.51}$$

Observe the quadratic exponent, which makes it different from the noncommutator case. As an easy consequence of the extrapolation theorem mentioned above, we have the following.

Corollary 3.37. *Let T be a linear operator satisfying (3.50) and let $1 < p < \infty$. Then there is a constant $c_{n,p}$ such that*

$$\|[b, T]\|_{L^p(w)} \leq c_{n,p} \, [w]_{A_p}^{\max\left\{1, \frac{1}{p-1}\right\}} \, \|b\|_{BMO}. \tag{3.52}$$

In [12], the special cases of the Hilbert, Beurling and Riesz transforms were obtained.

Sketch of proof of Theorem 3.36. We "conjugate" the operator as follows: If z is any complex number, we define

$$T_z(f) = e^{zb} \, T(e^{-zb} f).$$

Then, a computation gives (for "nice" functions):

$$[b, T](f) = \frac{d}{dz} T_z(f)|_{z=0} = \frac{1}{2\pi i} \int_{|z|=\epsilon} \frac{T_z(f)}{z^2} \, dz, \qquad \epsilon > 0,$$

by the Cauchy integral theorem; see [1, 16].

Now, by Minkowski's inequality,

$$\|[b, T](f)\|_{L^2(w)} \leq \frac{1}{2\pi\epsilon^2} \int_{|z|=\epsilon} \|T_z(f)\|_{L^2(w)} \, |dz|, \qquad \epsilon > 0. \tag{3.53}$$

The key point is to find the appropriate radius ϵ. To do this, we look at the inner norm $\|T_z(f)\|_{L^2(w)}$,

$$\|T_z(f)\|_{L^2(w)} = \|T(e^{-zb}f)\|_{L^2(we^{2\,\mathrm{Re}\,zb})},$$

and try to find appropriate bounds on z. To do this, we use the main hypothesis, namely that T is bounded on $L^2(w)$ if $w \in A_2$ with

$$\|T\|_{L^2(w)} \leq c\,[w]_{A_2}.$$

Hence we should estimate $[we^{2\,\mathrm{Re}\,zb}]_{A_2}$. Since $w \in A_2$, we can use Lemma 3.28 with $p = 2$ to obtain

$$[we^{2\,\mathrm{Re}\,zb}]_{A_2} \leq 4\,[w]_{A_2}\,[e^{2\,\mathrm{Re}\,zr'b}]_{A_2}^{1/r'},$$

where $r = r_w = 1 + \frac{1}{2^{n+5}[w]_{A_2}} < 2$. Now, since $b \in BMO$, we are in a position to apply Lemma 3.35: If

$$|2\,\mathrm{Re}\,zr'| \leq \frac{\alpha_n}{\|b\|_{BMO}},$$

then $[e^{2\,\mathrm{Re}\,zr'b}]_{A_2} \leq \beta_n^2$. Hence, for these z,

$$[we^{2\,\mathrm{Re}\,zb}]_{A_2} \leq 4\,[w]_{A_2}\,\beta_n^{2/r'} \leq 4\,[w]_{A_2}\,\beta_n,$$

since $1 < r < 2$.

Using this estimate and for these z we have

$$\|T_z(f)\|_{L^2(w)} \leq 4\beta_n\,[w]_{A_2}\,\|f\|_{L^2(w)}.$$

Finally, choosing the radius

$$\epsilon = \frac{\alpha_n}{2r'\|b\|_{BMO}}$$

finishes the proof of the theorem. All the details can be found in [13]. $\square$

3.10.3 Examples

Below there is an example illustrating that the quadratic estimate is sharp for $p = 2$ in dimension one.

Consider the Hilbert transform

$$Hf(x) = \text{p.v.} \int_{\mathbb{R}} \frac{f(y)}{x - y}\, dy,$$

and consider the BMO function $b(x) = \log|x|$ and

$$[b, H]f(x) = b(x)H(f)(x) - H(bf)(x).$$

Consider the BMO function $b(x) = \log|x|$. We know that

$$\|[b, H]\|_{L^2(w)} \le c\,[w]_{A_2}^2 \tag{3.54}$$

and we show that the result is sharp:

$$\sup_{w \in A_2} \frac{1}{[w]_2^{2-\theta}}\, \|[b, T]\|_{L^2(w)} = \infty, \qquad \theta > 0. \tag{3.55}$$

For $0 < \delta < 1$ we let

$$w(x) = |x|^{1-\delta}$$

and it is easy to see that

$$[w]_{A_2} \approx \frac{1}{\delta}.$$

We now consider the function

$$f(x) = x^{-1+\delta}\chi_{(0,1)}(x)$$

and observe that f is in $L^2(w)$.

To estimate $\|[b, H]f\|_{L^2(w)}$, we claim that

$$|[b, H]f(x)| \ge \frac{1}{\delta^2}\, f(x)$$

and hence

$$\|[b, H]f\|_{L^2(w)} \ge \frac{1}{\delta^2}\, \|f\|_{L^2(w)},$$

from which the sharpness (3.55) will follow.

We now prove the claim: If $0 < x < 1$, then

$$[b, H]f(x) = \int_0^1 \frac{\log(x) - \log(y)}{x - y}\, y^{-1+\delta}\, dy = \int_0^1 \frac{\log(\frac{x}{y})}{x - y}\, y^{-1+\delta}\, dy$$

$$= x^{-1+\delta} \int_0^{1/x} \frac{\log(\frac{1}{t})}{1 - t}\, t^{-1+\delta}\, dt.$$

Now,

$$\int_0^{1/x} \frac{\log(\frac{1}{t})}{1-t}\, t^{-1+\delta}\, dt = \int_0^1 \frac{\log(\frac{1}{t})}{1-t}\, t^{-1+\delta}\, dt + \int_1^{1/x} \frac{\log(\frac{1}{t})}{1-t}\, t^{-1+\delta}\, dt$$

and, since $\frac{\log(\frac{1}{t})}{1-t}$ is positive for $(0,1) \cup (1,\infty)$, we have, for $0 < x < 1$,

$$|[b,H]f(x)| > x^{-1+\delta} \int_0^1 \frac{\log(\frac{1}{t})}{1-t}\, t^{-1+\delta}\, dt.$$

But since

$$\int_0^1 \frac{\log(\frac{1}{t})}{1-t}\, t^{-1+\delta}\, dt > \int_0^1 \log\left(\frac{1}{t}\right) t^{-1+\delta}\, dt = \int_0^\infty s\, e^{-s\delta}\, ds = \frac{1}{\delta^2},$$

the claim

$$|[b,H]f(x)| \geq \frac{1}{\delta^2}\, f(x)$$

follows.

3.10.4 The A_1 case

In this section we describe some recent work by C. Ortiz-Caraballo [62]. The first result is a version of the linear growth Theorem 3.5 for commutators.

Theorem 3.38. *Let T be a Calderón–Zygmund operator and let b be in BMO. Also let $1 < p, r < \infty$. Then there exists a constant c_n such that, for any weight w, the following inequality holds if $1 < p, r < \infty$:*

$$\|b,T]f\|_{L^p(w)} \leq cp(p')^2\, \|b\|_{BMO}\, (r')^{1+1/p'}\, \|f\|_{L^p(M_r w)}. \tag{3.56}$$

In particular, if $w \in A_1$, we have

$$\|[b,T]\|_{L^p(w)} \leq c_n\, \|b\|_{BMO}\, p(p')^2 [w]_{A_1}^2. \tag{3.57}$$

Furthermore, this result is sharp in terms of the $[w]_{A_1}$ constant and in terms of p.

Observe again the quadratic exponent. We remit to [62] for the proof of the theorem.

The second main result of [61] is the following endpoint version of Theorem 3.38, similar in spirit as in the logarithmic growth (Theorem 3.6).

Theorem 3.39. *Let T and b as above. There exists a constant $c = c(n, \|b\|_{BMO})$ such that, for any weight $w \in A_1$ and $f \in L_c^\infty(\mathbb{R}^n)$,*

$$w(\{x \in \mathbb{R}^n : |[b,T]f(x)| > \lambda\}) \leq C\, \Phi([w]_{A_1})^2 \int_{\mathbb{R}^n} \Phi\left(\frac{|f(x)|}{\lambda}\right) w(x)\, dx, \tag{3.58}$$

where $\Phi(t) = t(1 + \log^+ t)$.

More recently, C. Ortiz-Caraballo has obtained in her Ph.D. dissertation [61] an improvement of these theorems in terms of mixed A_1-A_∞ norms for the commutator and for its iterations in the spirit of Theorems 3.33 and 3.14 above derived in [38]. For any $k \in \mathbb{N}$, the k-th iterated commutator T_b^k of a BMO function b and a Calderón–Zygmund operator T is defined by

$$T_b^k = [b, T_b^{k-1}].$$

Theorem 3.40. *Let T and b as above and let $1 < p, r < \infty$. Consider the higher-order commutators T_b^k, $k = 1, 2, \ldots$. Then there exists a constant $c = c_{n,T}$ such that for any weight w the following inequality holds:*

$$\|T_b^k f\|_{L^p(w)} \le c \, \|b\|_{BMO}^k \, (pp')^{k+1} \, (r')^{k+1/p'} \, \|f\|_{L^p(M_r w)}. \tag{3.59}$$

In particular, if $w \in A_1$, we have that

$$\|T_b^k\|_{L^p(w)} \le c \, \|b\|_{BMO}^k (pp')^{k+1} [w]_{A_1}^{1/p} [w]_{A_\infty}^{k+1/p'}.$$

Theorem 3.41. *Let T and b as above, and let $1 < p, r < \infty$. Then there exists a constant $c = c_{n,T}$ such that, for any w,*

$$w(\{x \in \mathbb{R}^n : |T_b^k f(x)| > \lambda\})$$
$$\le c \, (pp')^{(k+1)p} (r')^{(k+1)p-1} \int_{\mathbb{R}^n} \Phi\left(\|b\|_{BMO} \frac{|f|}{\lambda}\right) M_r w \, dx, \tag{3.60}$$

where $\Phi(t) = t(1 + \log^+ t)^k$. If $w \in A_1$, we obtain that

$$w(\{x \in \mathbb{R}^n : |T_b^k f(x)| > \lambda\}) \le c_n \beta \int_{\mathbb{R}^n} \Phi\left(\|b\|_{BMO} \frac{|f(x)|}{\lambda}\right) w(x) \, dx,$$

where $\beta = [w]_{A_1} [w]_{A_\infty}^k (1 + \log^+ [w]_{A_\infty})^{k+1}$ and $\Phi(t) = t(1 + \log^+ t)^k$.

The very first results of this type can be found in [65] (see also [68]).

These results reflect the idea that the higher is the commutator the more singular it is. Again, as above, a crucial role is played by the sharp reverse Hölder property for A_∞ weights as stated in Theorem 3.31. See [61] for details.

3.11 Rearrangement type estimates

The main purpose of this section is to give a new proof of (3.39) avoiding any sort of good-λ type arguments like (3.40) used in [52] and [54]. Indeed, as already explained in Section 3.4.2, the heart of the matter of the proof of Theorem 3.5 and hence of Theorem 3.6 is Lemma 3.22. We plan to give a proof based on the idea of rearrangements.

Recall that the nonincreasing rearrangement f^* of a measurable function f is defined by

$$f^*(t) = \inf \{\lambda > 0 : \mu_f(\lambda) < t\}, \quad t > 0,$$

where

$$\mu_f(\lambda) = |\{x \in \mathbb{R}^n : |f(x)| > \lambda\}|, \quad \lambda > 0,$$

is the distribution function of f. More generally, for a weight w (or measure) we define

$$f_w^*(t) = \inf \{\lambda > 0 : w_f(\lambda) < t\}, \quad t > 0,$$

where

$$w_f(\lambda) = w(\{x \in \mathbb{R}^n : |f(x)| > \lambda\}), \quad \lambda > 0.$$

An important fact is that

$$\int_{\mathbb{R}^n} |f|^p\, w\, dx = \int_0^\infty f_w^*(t)^p\, dt,$$

or more generally, if E is any measurable function,

$$\int_E |f|^p\, w\, dx = \int_0^{w(E)} f^*(t)^p\, dt.$$

Similarly,

$$\|f\|_{L^{p,\infty}(w)} = \sup_{t>0} tw(\{x \in \mathbb{R}^n : |f(x)| > t\})^{1/p} = \sup_{t>0} t^{1/p} f_w^*(t).$$

If f is only a *measurable* function and Q is a cube, then we define the following quantity:

$$(f\chi_Q)^*(\lambda|Q|), \qquad 0 < \lambda \le 1.$$

We can think of this expression as another way of averaging the function. Indeed, for any $\delta > 0$ and $0 < \lambda \le 1$,

$$(f\chi_Q)^*(\lambda|Q|) \le \left(\frac{1}{\lambda|Q|} \int_Q |f|^\delta\, dx\right)^{1/\delta}.$$

The (dyadic) local maximal operator $m_\lambda f$ is defined for any measurable function f by

$$m_\lambda f(x) = \sup_{x \in Q \in \mathcal{D}} (f\chi_Q)^*(\lambda|Q|), \qquad 0 < \lambda < 1,$$

where f^* denotes the nonincreasing rearrangement of f.

We may use the following property:

$$|f(x)| \le m_\lambda f(x), \quad \text{a.e. } x; \tag{3.61}$$

see [46, Lemma 6].

Given a cube Q, define the *median value* $m_f(Q)$ of f over Q as a (possibly nonunique) number such that

$$|\{x \in Q : |f(x)| > m_f(Q)\}| \leq |Q|/2$$

and

$$|\{x \in Q : |f(x)| < m_f(Q)\}| \leq |Q|/2.$$

It follows easily from the definition that

$$|m_f(Q)| \leq (f\chi_Q)^*(|Q|/2).$$

Also, it is easy to see that, if f is nonnegative,

$$m_f(Q) = (f\chi_Q)^*(|Q|/2)$$

and, for any constant c,

$$m_f(Q) - c = m_{f-c}(Q), \tag{3.62}$$

which in turns implies that f is positive and $\delta > 0$.

We will use

$$|m_f(P) - m_f(Q)| = |m_{f-m_f(Q)}(P)|$$
$$\leq ((f - m_f(Q))\chi_P)^*(|P|/2) \tag{3.63}$$
$$\leq \left(\frac{2}{|P|} \int_P |f - m_f(Q)|^\delta \, dx\right)^{1/\delta}.$$

Theorem 3.42. *Let $w \in A_q$ and let $0 < \delta < 1$ and $0 < \gamma < 1$. There is a constant $c = c_{n,q,\gamma,\delta}$ such that, for any measurable function f,*

$$f_w^*(t) \leq c\,[w]_{A_q}\,(M_\delta^\# f)_w^*(\gamma\,t) + f_w^*(2t), \qquad t > 0. \tag{3.64}$$

The term $f_w^*(2t)$ is a sort of "error" term. This kind of estimates goes back to the work of R. Bagby and D. Kurtz in the mid eighties [4, 5].

We recall the sharp maximal operator: If $\delta > 0$, then

$$M_\delta^\# f(x) = M^\#(|f|^\delta)(x)^{1/\delta},$$

where $M^\#$ is the usual sharp maximal function of Fefferman–Stein:

$$M^\#(f)(x) = \sup_{Q \ni x} \frac{1}{|Q|} \int_Q |f(y) - f_Q| \, dy,$$

and where $f_Q = \frac{1}{|Q|} \int_Q f(y) \, dy$. An interesting observation is that the function can be non locally integrable in the interesting case $0 < \delta < 1$.

If we iterate (3.64), we have:

$$f_w^*(t) \leq c\,[w]_{A_q} \sum_{k=0}^{\infty} (M_\delta^\# f)_w^*(2^k \gamma t) + f_w^*(+\infty)$$

$$\leq \frac{[w]_{A_q}}{\log 2} \int_{t\gamma/2}^{\infty} (M_\delta^\# f)_w^*(s)\,\frac{ds}{s} + f_w^*(+\infty).$$

Hence, if we assume that

$$f_w^*(+\infty) = 0,$$

the inequality that we obtain is

$$f_w^*(t) \leq c\,[w]_{A_q} \int_{t\gamma/2}^{\infty} (M_\delta^\# f)_w^*(s)\,\frac{ds}{s}. \tag{3.65}$$

We continue using the Hardy operator. Recall that if $f\colon (0,\infty) \to [0,\infty)$ then

$$Af(x) = \frac{1}{x} \int_0^x f(t)\,dt, \qquad x > 0,$$

is called the *Hardy operator*. The dual operator is given by

$$Sf(x) = \int_x^\infty f(s)\frac{ds}{s}.$$

Hence, the above estimate can be expressed as

$$f_w^*(t) \leq c\,[w]_{A_q}\, S((M_\delta^\# f)_w^*)(t\gamma/2).$$

Finally, since it is well known that these operators are bounded on $L^p(0,\infty)$ and, furthermore, it is known that

$$\|S\|_{L^p(0,\infty)} = p, \qquad p \geq 1,$$

we have

$$\|f\|_{L^p(w)} = \|f_w^*\|_{L^p(0,\infty)} \leq c\,[w]_{A_q}\, \|S((M_\delta^\# f)_w^*)\|_{L^p(0,\infty)}$$

$$\leq cp\,[w]_{A_q}\, \|(M_\delta^\# f)_w^*\|_{L^p(0,\infty)} = cp\,[w]_{A_q}\, \|M_\delta^\# f\|_{L^p(w)}.$$

This concludes the strong estimate in the case $p \leq 1$. The triangle inequality can be used when $0 < p < 1$ and the weak estimate is easier. This concludes the proof of the following corollary, except for the proof of Theorem 3.42.

Corollary 3.43. *Let $0 < p < \infty$ and $w \in A_q$. Suppose that f is such that*

$$|\{x : f(x) > t\}| < \infty, \qquad t > 0. \tag{3.66}$$

Then

$$\|f\|_{L^p(w)} \leq cp\,[w]_{A_q}\|M_\delta^\# f\|_{L^p(w)} \tag{3.67}$$

and

$$\|f\|_{L^{p,\infty}(w)} \leq cp\,[w]_{A_q}\|M_\delta^\# f\|_{L^{p,\infty}(w)}. \tag{3.68}$$

These estimates have been improved in [63], where $[w]_{A_q}$ was replaced by $[w]_{A_\infty}$. As before, this is Wilson's constant (3.47).

To understand the reason of assuming (3.66), we observe that $f_w^*(+\infty) = 0$ is equivalent to saying that, for each $t > 0$,

$$w(\{x \, : \, |f(x)| > t\}) < \infty.$$

This condition was already used in [5] (see also [49]). Now, since $w \in A_q$ and since $\min\{w, N\} \in A_q$, $N = 1, 2, \ldots$, with A_q constant and independent of N, we may infer that w is bounded. This explains why it is enough to consider functions satisfying (3.66).

Finally, combining this corollary with Lemma 3.23, we derive the sharp Coifman–Fefferman inequality in both p and the A_q constant, which was needed for the proof of the main results in these lecture notes.

Corollary 3.44. *Let $0 < p < \infty$ and $w \in A_q$. Let T be any Calderón–Zygmund operator. Suppose that f is a smooth function such that, for each $t > 0$, we have $|\{x \, : \, |Tf(x)| > t\}| < \infty$. Then*

$$\|Tf\|_{L^p(w)} \leq cp \, [w]_{A_q} \|Mf\|_{L^p(w)} \tag{3.69}$$

and

$$\|Tf\|_{L^{p,\infty}(w)} \leq cp \, [w]_{A_q} \|Mf\|_{L^{p,\infty}(w)}. \tag{3.70}$$

We note that (3.69) was already pointed out in (3.24).

3.12   Proof of Theorem 3.42

Proof. As already explained in the previous section, we may assume that the weight is bounded. We may also assume that f is nonnegative. Fix $t > 0$ and let $A = [w]_{A_q}$. By definition of rearrangement, (3.64) will follow if we prove that, for each $t > 0$,

$$w(\{x \in \mathbb{R}^n \, : \, f(x) > cA \left(M_\delta^\# f\right)_w^* (\gamma t) + f_w^*(2t)\}) \leq t. \tag{3.71}$$

We split the left-hand side L as follows:

$$L \leq w(\{x \in \mathbb{R}^n \, : \, cA \, M_\delta^\# f(x) > cA \left(M_\delta^\# f\right)_w^* (\gamma t)\})$$

$$+ w(\{x \in \mathbb{R}^n \, : \, f(x) > cA \, M_\delta^\# f(x) + f_w^*(2t)\}) = I + II.$$

Observe that

$$I = w(\{x \in \mathbb{R}^n \, : \, M_\delta^\# f(x) > \left(M_\delta^\# f\right)_w^* (\gamma \, t)\}) \leq \gamma t,$$

by definition of rearrangement, and hence the heart of the matter is to prove that

$$II = w(\{x \in \mathbb{R}^n \, : \, f(x) > cA \, M_\delta^\# f(x) + f_w^*(2t)\}) \leq (1 - \gamma) \, t.$$

The set $\{x \in \mathbb{R}^n \, : \, f(x) > cA \, M_\delta^\# f(x) + f_w^*(2t)\}$ is contained in the set $E = \{x \in \mathbb{R}^n \, : \, f(x) > f_w^*(2t)\}$, which has w-measure at most $2t$ by definition of rearrangement. Now, by the regularity of the measure, we can find an open set Ω containing E such that $w(\Omega) < 3t$. We claim that, for a large dimensional constant c,

$$II = w(\{x \in \Omega \,:\, f(x) > cA\, M_\delta^\# f(x) + f_w^*(2t)\}) \leq (1 - \gamma)\, t.$$

Since we may assume that $|\Omega|$ is also finite, we can consider the Calderón–Zygmund cubes of the function χ_Ω at level $0 < \alpha < 1$. Hence, there are dyadic cubes $\{Q_j\}$, maximal with respect to inclusion, satisfying

$$\Omega \subset \cup_j Q_j$$

and

$$\alpha < \frac{|\Omega \cap Q_j|}{|Q_j|} \leq 2^n \alpha \tag{3.72}$$

for each j. Observe that if we further choose α such that $2^n \alpha < \frac{1}{2}$ we have the following important property:

$$\frac{|\Omega^c \cap Q_j|}{|Q_j|} \geq \frac{1}{2}. \tag{3.73}$$

Now, we continue estimating II:

$$II \leq \sum_j w(\{x \in Q_j \,:\, f(x) > cA\, M_\delta^\# f(x) + f_w^*(2t)\}).$$

Fix one of these j. If $|m_f(Q_j)| = m_f(Q_j) \leq f_w^*(2t)$, we have (recall that $m_f(Q)$ is the median value of f) that

$$\{x \in Q_j \,:\, f(x) > cA\, M_\delta^\# f(x) + f_w^*(2t)\}$$

$$\subset \{x \in Q_j \,:\, |f(x) - m_f(Q_j)| > cA\, M_\delta^\# f(x)\}.$$

Hence, assuming $m_f(Q_j) > f_w^*(2t)$ and since $f(x) \leq f_w^*(2t)$ if $x \in \Omega^c$, we have

$$\frac{1}{|Q_j|} \int_{Q_j} |f(y) - m_f(Q_j)|^\delta \, dy \geq \frac{1}{|Q_j|} \int_{\Omega^c \cap Q_j} \Big| |m_f(Q_j)| - |f(y)| \Big|^\delta \, dy$$

$$\geq \frac{|\Omega^c \cap Q_j|}{|Q_j|} \Big(|m_f(Q_j)| - f_w^*(2t) \Big)^\delta \geq \frac{1}{2} \Big(|m_f(Q_j)| - f_w^*(2t) \Big)^\delta.$$

Hence,

$$|m_f(Q_j)| \leq |m_f(Q_j) - f_w^*(2t)| + f_w^*(2t)$$

$$\leq \left(\frac{2}{|Q_j|} \int_{Q_j} |f(y) - m_f(Q_j)|^\delta \, dy \right)^{1/\delta} + f_w^*(2t)$$

$$\leq 2^{1/\delta} M_\delta^\# f(x) + f_w^*(2t)$$

if $x \in Q_j$. Combining estimates we have that, if $c > 1 + 2^{1/\delta}$,

$$\{x \in Q_j : f(x) > cA\,M_\delta^\# f(x) + f_w^*(2t)\}$$

$$\subset \{x \in Q_j : |f(x) - m_f(Q_j)| > cA\,M_\delta^\# f(x)\}$$

for any j. Now, we denote

$$E_j = \{x \in Q_j : |f(x) - m_f(Q_j)| > cA\,M_\delta^\# f(x)\} \qquad (3.74)$$

and observe that, by Lemma 3.45 below, we have

$$\frac{|E_j|}{|Q_j|} \le c_1 e^{-c_2 A}$$

since $A = [w]_{A_q}$. To conclude, if $r = r_w$ is the sharp reverse exponent for w, i.e., $r_w = 1 + \frac{1}{2^{2q+n+1}A}$ (see Lemma 3.28),

$$\left(\frac{1}{|Q|}\int_Q w^{r_w}\,dx\right)^{\frac{1}{r_w}} \le \frac{2}{|Q|}\int_Q w,$$

we have, by Hölder's inequality and Lemma 3.45,

$$\frac{w(E_j)}{w(Q_j)} \le 2\left(\frac{|E_j|}{|Q_j|}\right)^{1/r'} \le \frac{2c_1}{e^{cc_3}},$$

since $r' \approx [w]_{A_q} = A$, where c_3 depends on the dimension and q. If we finally choose c such that $\frac{2c_1}{e^{cc_3}} < \frac{1-\gamma}{3}$, we have

$$II \le \sum_j w(E_j) \le \frac{1-\gamma}{3}\sum_j w(Q_j) = \frac{1-\gamma}{3}\,w(\Omega) < (1-\gamma)\,t,$$

since $w(\Omega) < 3t$.   □

3.13   The exponential decay lemma

Let $f \in L^\delta_{\mathrm{loc}}$ be measurable. For $t > 0$, we define

$$\varphi(t) = \sup_{Q \in \mathcal{D}} \frac{1}{|Q|}\left|\{x \in Q : |f(x) - m_f(Q)| > t\,M_\delta^\# f(x)\}\right|,$$

which is finite since it is bounded by 1 and by $1/t^\delta$, by Chebyshev. We prove that there is an exponential decay on t.

Lemma 3.45. *There are dimensional constants c_1, c_2 such that*

$$\varphi(t) \le \frac{c_1}{e^{c_2 t}}. \qquad (3.75)$$

Proof. Observe that we may assume that f is positive by the lattice property. We denote

$$\mathrm{osc}_\delta(Q, f) = \left(\frac{1}{|Q|} \int_Q |f(y) - m_f(Q)|^\delta \, dy \right)^{1/\delta}.$$

It is enough to find $t_0 > 0$ such that $\varphi(t) \leq \frac{c}{e^{ct}}$ for $t > t_0$. Fix one cube Q and consider the ratio

$$\frac{1}{|Q|} \left| \{ x \in Q : |f(x) - m_f(Q)| > t\, M_\delta^\# f(x) \} \right|. \tag{3.76}$$

Consider the maximal type operator related to Q,

$$\mathcal{N}_Q f(x) = \sup_{x \in P \in \mathcal{D}(Q)} |m_f(P) - m_f(Q)|,$$

and consider the subset of Q

$$\Gamma = \{ x \in Q : \mathcal{N}_Q f(x) > \sigma\, \mathrm{osc}_\delta(Q, f) \},$$

where the parameter $\sigma > 1$ is going to be chosen. Observe that if Γ is empty then using that $m_f(P) = (f \chi_P)^*(|P|/2)$ when f is positive, by the Lebesgue differentiation theorem, shrinking P to x we have

$$|f(x) - m_f(Q)| \leq \sigma\, \mathrm{osc}_\delta(Q, f) \leq \sigma\, M_\delta^\# f(x), \quad x \in Q,$$

so that if $t > \sigma$ then the ratio (3.76) is zero. So we will assume that $t > \sigma$ and that Γ is not empty. Hence, by adapting the usual CZ covering lemma there is a collection $\{Q_i\}$ of dyadic disjoint *subcubes* of Q with the usual properties:

$$\Gamma = \cup_i Q_i$$

and each is maximal with respect to

$$\sigma\, \mathrm{osc}_\delta(Q, f) < |m_f(Q_i) - m_f(Q)|. \tag{3.77}$$

Hence,

$$|m_f(Q_i') - m_f(Q)| \leq \sigma\, \mathrm{osc}_\delta(Q, f). \tag{3.78}$$

Observe that, as before,

$$|f(x) - m_f(Q)| \leq \sigma\, \mathrm{osc}_\delta(Q, f), \quad x \in Q \setminus \Omega. \tag{3.79}$$

Observe also that, by (3.63) and (3.77),

$$|\Gamma| \leq \frac{2}{\sigma^\delta}. \tag{3.80}$$

Combining and assuming that $t > \sigma$,

$$\left| \{ x \in Q : |f(x) - m_f(Q)| > t\, M_\delta^\# f(x) \} \right|$$

$$= \left| \{ x \in \Omega : |f(x) - m_f(Q)| > t\, M_\delta^\# f(x) \} \right|$$

$$= \sum_i \left| \{ x \in Q_i : |f(x) - m_f(Q)| > t M_\delta^\# f(x) \} \right|.$$

By (3.78),

$$|m_f(Q_i') - m_f(Q)| \leq \sigma \operatorname{osc}_\delta(Q, f) \leq \sigma M_\delta^\# f(x)$$

$$\leq \sum_i \left| \{x \in Q_i \,:\, |f(x) - m_f(Q_i')| > (t - \sigma) M_\delta^\# f(x)\} \right|.$$

Now, recalling that, by definition, if $t > 0$,

$$\varphi(t) = \sup_Q \frac{1}{|Q|} \left| \{x \in Q \,:\, |f(x) - m_f(Q)| > t M^\# f(x)\} \right|$$

and hence, if $t > \sigma$,

$$\frac{1}{|Q|} \left| \{x \in Q \,:\, |f(x) - m_f(Q)| > t M_\delta^\# f(x)\} \right| \leq \frac{1}{|Q|} \sum_i |Q_i'| \, \varphi(t - \sigma)$$

$$\leq \frac{2^{n+1}}{\sigma^\delta} \, \varphi(t - \sigma),$$

by (3.80). Hence, if we choose σ such that $\sigma^\delta = 2^{n+1}e$, then, for any $t > \sigma$, we have

$$\varphi(t) \leq \frac{1}{e} \, \varphi(t - \sigma).$$

Reiterating, we get, for $k = 1, 2, \ldots$,

$$\varphi(t) \leq \frac{1}{e^k} \, \varphi(t - \sigma k),$$

until k gets exhausted, namely $k \approx \frac{t}{\sigma}$. Thus we obtain

$$\varphi(t) \leq \frac{c_1}{e^{c_2 t}}.$$

(The case $t \leq \sigma$ is trivial.) $\square$

Bibliography

[1] J. Álvarez, R. Bagby, D. Kurtz and C. Pérez, *Weighted estimates for commutators of linear operators*, Studia Math. **104** (1993), no. 2, 195–209.

[2] J. Álvarez and C. Pérez, *Estimates with A_∞ weights for various singular integral operators*, Bollettino U.M.I. (7) 8-A (1994), 123–133.

[3] K. Astala, T. Iwaniec and E. Saksman, *Beltrami operators in the plane*, Duke Math. J. **107** (2001), no. 1, 27–56.

[4] R. J. Bagby and D. S. Kurtz, *Covering lemmas and the sharp function*, Proc. Amer. Math. Soc. **93** (1985), 291–296.

[5] R. J. Bagby and D. S. Kurtz, *A rearranged good-λ inequality*, Trans. Amer. Math. Soc. **293** (1986), 71–81.

[6] O. Beznosova, *Linear bound for dyadic paraproduct on weighted Lebesgue space $L^2(w)$*, J. Funct. Anal. **255** (2008), no. 4, 994–1007.

[7] O. Beznosova and A. Reznikov, *Sharp estimates involving A_∞ and $L \log L$ constants, and their applications to PDE*, preprint (2011), arXiv:1107.1885.

[8] S. M. Buckley, *Estimates for operator norms on weighted spaces and reverse Jensen inequalities*, Trans. Amer. Math. Soc. **340** (1993), no. 1, 253–272.

[9] M. J. Carro, C. Pérez, F. Soria and J. Soria, *Examples and counterexamples for fractional operators*, Indiana Univ. Math. J. **54** (2005), 627–644.

[10] S.-Y. A. Chang, J. M. Wilson and T. Wolff, *Some weighted norm inequalities concerning the Schrödinger operator*, Comment. Math. Helv. **60** (1985), 217–246.

[11] S. Chanillo and R. L. Wheeden, *Some weighted norm inequalities for the area integral*, Indiana Univ. Math. J. **36** (1987), 277–294.

[12] D. Chung, *Sharp estimates for the commutators of the Hilbert, Riesz transforms and the Beurling–Ahlfors operator on weighted Lebesgue spaces*, preprint (2010), arXiv:1001.0755, Indiana Univ. Math. J. (to appear).

[13] D. Chung, M. C. Pereyra and C. Pérez, *Sharp bounds for general commutators on weighted Lebesgue spaces*, Trans. Amer. Math. Soc. **364** (2012), 1163–1177.

[14] R. R. Coifman and C. Fefferman, *Weighted norm inequalities for maximal functions and singular integrals*, Studia Math. **51** (1974), 241–250.

[15] R. R. Coifman and R. Rochberg, *Another characterization of BMO*, Proc. Amer. Math. Soc. **79** (1980), 249–254.

[16] R. Coifman, R. Rochberg and G. Weiss, *Factorization theorems for Hardy spaces in several variables*, Ann. of Math. **103** (1976), 611–635.

[17] A. Córdoba and C. Fefferman, *A weighted norm inequality for singular integrals*, Studia Math. **57** (1976), 97–101.

[18] D. Cruz-Uribe, J. M. Martell and C. Pérez, *Extrapolation results for A_∞ weights and applications*, J. Funct. Anal. **213** (2004), no. 2, 412–439.

[19] D. Cruz-Uribe, SFO, J. M. Martell and C. Pérez, *Weighted weak-type inequalities and a conjecture of Sawyer*, Int. Math. Res. Not. **30** (2005), 1849–1871.

[20] D. Cruz-Uribe, SFO, J. M. Martell and C. Pérez, *Weights, Extrapolation and the Theory of Rubio de Francia*, Operator Theory: Advances and Applications, vol. 215, Birkhäuser, Basel, 2011,
http://www.springer.com/mathematics/analysis/book/978-3-0348-0071-6

[21] D. Cruz-Uribe, SFO, J. M. Martell and C. Pérez, *Sharp weighted estimates for classical operators*, Adv. Math. **229** (2012), 408–441.

[22] D. Cruz-Uribe, SFO, J. M. Martell and C. Pérez, *Sharp weighted estimates for approximating dyadic operators*, Electron. Res. Announc. Math. Sci. **17** (2010), 12–19.

[23] D. Cruz-Uribe and C. Pérez, *Two weight extrapolation via the maximal operator*, J. Funct. Anal. **174** (2000), no. 1, 1–17.

[24] G. P. Curbera, J. García-Cuerva, J. M. Martell and C. Pérez, *Extrapolation with weights, rearrangement invariant function spaces, modular inequalities and applications to singular integrals*, Adv. Math. **203** (2006), 256–318.

[25] M. de Guzmán, *Differentiation of Integrals in $\mathbb{R}^n$*, Lectures Notes in Math., vol. 481, Springer-Verlag, Berlin, Heidelberg, New York, 1975.

[26] O. Dragičević, L. Grafakos, M. C. Pereyra and S. Petermichl, *Extrapolation and sharp norm estimates for classical operators on weighted Lebesgue spaces*, Publ. Math. **49** (2005), no. 1, 73–91.

[27] J. Duoandikoetxea, *Fourier Analysis*, Amer. Math. Soc. Grad. Stud. Math., vol. 29, Providence, RI, 2000.

[28] R. Fefferman and J. Pipher, *Multiparameter operators and sharp weighted inequalities*, Amer. J. Math. **119** (1997), no. 2, 337–369.

[29] C. Fefferman and E. M. Stein, *Some maximal inequalities*, Amer. J. Math. **93** (1971), 107–115.

[30] N. Fujii, *A proof of the Fefferman-Stein-Strömberg inequality for the sharp maximal functions*, Proc. Amer. Math. Soc. **106** (1989), no. 2, 371–377.

[31] J. García-Cuerva and J. L. Rubio de Francia, *Weighted Norm Inequalities and Related Topics*, North-Holland Math. Studies, vol. 116, North-Holland, Amsterdam, 1985.

[32] L. Grafakos, *Classical Fourier Analysis*, Graduate Texts in Math., vol. 249, 2nd Ed., Springer-Verlag, New York, 2008.

[33] L. Grafakos, *Modern Fourier Analysis*, Graduate Texts in Math., vol. 250, 2nd Ed., Springer-Verlag, New York, 2008.

[34] E. Hernández, *Factorization and extrapolization of pairs of weights*, Studia Math. **95** (1989), 179–193.

[35] S. Hruščev, *A description of weights satisfying the A_∞ condition of Muckenhoupt*, Proc. Amer. Math. Soc. **90** (1984), no. 2, 253–257.

[36] T. Hytönen, *The sharp weighted bound for general Calderón–Zygmund operators*, Ann. of Math. **175** (2012), no. 3, 1473–1506.

[37] T. Hytönen and M. T. Lacey, *The A_p-A_∞ inequality for general Calderón–Zygmund operators*, Indiana Univ. Math. J. (to appear).

[38] T. Hytönen and C. Pérez, *Sharp weighted bounds involving A_∞*, Anal. Partial Differential Equations (to appear).

[39] T. Hytönen and C. Pérez, work in progress, 2012.

[40] T. Hytönen, C. Pérez, S. Treil and A. Volberg, *Sharp weighted estimates for dyadic shifts and the A_2 conjecture*, preprint (2010), arXiv:1010.0755.

[41] J. L. Journé, *Calderón–Zygmund operators, pseudo–differential operators and the Cauchy integral of Calderón*, Lecture Notes in Math., vol. 994, Springer-Verlag, Berlin, Heidelberg, New York, 1983.

[42] M. Lacey, *An A_p-A_∞ inequality for the Hilbert transform*, preprint (2011), arXiv:1104.2199, Houston J. Math. (to appear).

[43] M. Lacey, On the A_2 inequality for Calderón–Zygmund operators, preprint (2011), arXiv:1106.4802.

[44] M. Lacey, K. Moen, C. Pérez and R. Torres, *The sharp bound the fractional operators on weighted L^p spaces and related Sobolev inequalities*, J. Funct. Anal. **259** (2010), 1073–1097.

[45] M. Lacey, S. Petermichl and M. C. Reguera, *Sharp A_2 inequality for Haar shift operators*, Math. Ann. **348** (2010), no. 1, 127–141.

[46] A. K. Lerner, *On some pointwise inequalities*, J. Math. Anal. Appl. **289** (2004), no. 1, 248–259.

[47] A. K. Lerner, *On some sharp weighted norm inequalities*, J. Funct. Anal. **232** (2006), 477–494.

[48] A. K. Lerner, *An elementary approach to several results on the Hardy–Littlewood maximal operator*, Proc. Amer. Math. Soc. **136** (2008), no. 8, 2829–2833.

[49] A. K. Lerner, *Weighted rearrangement inequalities for local sharp maximal functions*, Trans. Amer. Math. Soc. **357** (2004), 2445–2465.

[50] A. K. Lerner, *A pointwise estimate for local sharp maximal function with applications to singular integrals*, Bull. Lond. Math. Soc. **42** (2010), no. 5, 843–856.

[51] A. K. Lerner, *Sharp weighted norm inequalities for Littlewood–Paley operators and singular integrals*, Adv. Math. **226** (2011), no. 5, 3912–3926.

[52] A. K. Lerner, S. Ombrosi and C. Pérez, *Sharp A_1 bounds for Calderón–Zygmund operators and the relationship with a problem of Muckenhoupt and Wheeden*, Int. Math. Res. Not. **2008**, article ID rnm161, 11 pp.

[53] A. Lerner, S. Ombrosi and C. Pérez, *Weak type estimates for Singular Integrals related to a dual problem of Muckenhoupt–Wheeden*, J. Fourier Anal. Appl. **15** (2009), no. 3, 394–403.

[54] A. Lerner, S. Ombrosi and C. Pérez, A_1 *bounds for Calderón–Zygmund operators related to a problem of Muckenhoupt and Wheeden*, Math. Res. Lett. **16** (2009), 149–156.

[55] J. M. Martell, C. Pérez and R. Trujillo-González, *Lack of natural weighted estimates for some singular integral operators*, Trans. Amer. Math. Soc. **357** (2005), 385–396.

[56] B. Muckenhoupt, *Weighted norm inequalities for the Hardy–Littlewood maximal function*, Trans. Amer. Math. Soc. **165** (1972), 207–226.

[57] B. Muckenhoupt and R. Wheeden, *Weighted norm inequalities for fractional integrals*, Trans. Amer. Math. Soc. **192** (1974), 261–274.

[58] F. Nazarov, A. Reznikov, V. Vasyunin and A. Volberg, A_1 *conjecture: weak norm estimates of weighted singular operators and Bellman functions*, preprint (2010), http://sashavolberg.wordpress.com

[59] F. Nazarov, S. Treil and A. Volberg, *Two weight inequalities for individual Haar multipliers and other well localized operators*, Math. Res. Lett. **15** no. 3 (2008), 583–597.

[60] F. Nazarov and A. Volberg, *Bellman function, polynomial estimates of weighted dyadic shifts, and A_2 conjecture*, preprint (2011).

[61] C. Ortiz-Caraballo, Ph.D. thesis, Universidad de Sevilla, October 2011.

[62] C. Ortiz-Caraballo, *Quadratic A_1 bounds for commutators of singular integrals with BMO functions*, Indiana Univ. Math. J. (to appear).

[63] C. Ortiz-Caraballo, C. Pérez and E. Rela, *Improving bounds for singular operators via sharp reverse Hölder inequality for A_∞*, preprint (2011).

[64] C. Pérez, *Weighted norm inequalities for singular integral operators*, J. London Math. Soc. **49** (1994), 296–308.

[65] C. Pérez, *Endpoint estimates for commutators of singular integral operators*, J. Funct. Anal. **128** (1995), no. 1, 163–185.

[66] C. Pérez, *Sharp estimates for commutators of singular integrals via iterations of the Hardy–Littlewood maximal function*, J. Fourier Anal. Appl. **3** (1997), 743–756.

[67] C. Pérez, *Sharp weighted inequalities for the vector-valued maximal function*, Trans. Amer. Math. Soc. **352** (2000), 3265–3288.

[68] C. Pérez and G. Pradolini, *Sharp weighted endpoint estimates for commutators of singular integrals*, Michigan Math. J. **49** (2001), 23–37.

[69] C. Pérez, S. Treil and A. Volberg, *On A_2 conjecture and corona decomposition of weights*, preprint (2010), arXiv:1006.2630.

[70] S. Petermichl, *The sharp bound for the Hilbert transform on weighted Lebesgue spaces in terms of the classical A_p-characteristic*, Amer. J. Math. **129** (2007), no. 5, 1355–1375.

[71] S. Petermichl, *The sharp weighted bound for the Riesz transforms*, Proc. Amer. Math. Soc. **136** (2008), no. 4, 1237–1249.

[72] S. Petermichl and A. Volberg, *Heating of the Ahlfors–Beurling operator: weakly quasiregular maps on the plane are quasiregular*, Duke Math. J. **112** (2002), no. 2, 281–305.

[73] M. C. Reguera, Ph.D. thesis, Georgia Institute of Technology, Spring 2011.

[74] M. C. Reguera, *On Muckenhoupt–Wheeden conjecture*, Adv. Math. **227** (2011), 1436–1450.

[75] M. C. Reguera and C. Thiele, *The Hilbert transform does not map $L^1(Mw)$ to $L^{1,\infty}(w)$*, Math. Res. Lett. **19** (2012), no. 1, 1–7.

[76] S. Treil, *Sharp A_2 estimates of Haar shifts via Bellman function*, preprint (2011), arXiv:1105.2252.

[77] A. Vagharshakyan, *Recovering singular integrals from Haar shifts*, Proc. Amer. Math. Soc. **138** (2010), no. 12, 4303–4309.

[78] J. M. Wilson, *Weighted inequalities for the dyadic square function without dyadic A_∞*, Duke Math. J. **55** (1987), no. 1, 19–50.

[79] J. M. Wilson, *Weighted inequalities for the continuous square function*, Trans. Amer. Math. Soc. **314** (1989), no. 2, 661–692.

[80] J. M. Wilson, *Weighted Littlewood–Paley theory and exponential-square integrability*, Lecture Notes in Math., vol. 1924, Springer-Verlag, Berlin, Heidelberg, New York, 2008.

Chapter 4

De Giorgi–Nash–Moser Theory

Xiao Zhong[1]

4.1 Introduction

4.1.1 Equations

We consider the second-order, linear, elliptic equations with divergence structure

$$\operatorname{div}(\mathbb{A}(x)\nabla u(x)) = \sum_{i,j=1}^{n} \partial_{x_i}(a_{ij}(x)\,\partial_{x_j} u(x)) = 0. \tag{4.1}$$

Here $\mathbb{A}(x) = [a_{ij}(x)]_{i,j=1,2,\dots,n}$ is an $n \times n$ symmetric matrix with measurable entries $a_{ij}(x)$, defined in a domain $\Omega \subset \mathbb{R}^n$, $n \geq 2$. We assume the following ellipticity and boundedness conditions:

$$\lambda|\xi|^2 \leqslant \langle \mathbb{A}(x)\xi,\, \xi \rangle = \sum_{i,j=1}^{n} a_{ij}(x)\,\xi_i \xi_j \leqslant \Lambda|\xi|^2 \tag{4.2}$$

for all $\xi = (\xi_1, \dots, \xi_n) \in \mathbb{R}^n$ and for almost every $x \in \Omega$. Here $0 < \lambda \leq \Lambda < \infty$ are constants.

Note that, since $\mathbb{A}(x)$ is a symmetric matrix, it has n real eigenvalues $\lambda_i(x)$, $i = 1, \dots, n$. Condition (4.2) is equivalent to

$$\lambda \leq \lambda_i(x) \leq \Lambda$$

for all $i = 1, \dots, n$ and almost every $x \in \Omega$ (exercise).

---

[1]Supported by the Academy of Finland, project 127639.

Example 4.1. Let $\mathbb{A}(x)$ be the identity matrix. Then the condition (4.2) is true with $\lambda = \Lambda = 1$, and equation (4.1) is reduced to the Laplace equation

$$\Delta u = \sum_{i=1}^{n} \partial_{x_i} \partial_{x_i} u = 0.$$

Example 4.2. Let α be a constant, $0 < \alpha < 1$. Define $\mathbb{A}(x)$ in $\mathbb{R}^2$ as

$$\mathbb{A}(x) = \begin{pmatrix} \dfrac{x_1^2 + \alpha^2 x_2^2}{|x|^2} & (1 - \alpha^2) \dfrac{x_1 x_2}{|x|^2} \\[3ex] (1 - \alpha^2) \dfrac{x_1 x_2}{|x|^2} & \dfrac{\alpha^2 x_1^2 + x_2^2}{|x|^2} \end{pmatrix}, \quad x = (x_1, x_2).$$

Then we have (exercise)

$$\alpha^2 |\xi|^2 \leq \langle \mathbb{A}(x)\xi,\, \xi \rangle \leq |\xi|^2, \quad \forall x \in \mathbb{R}^2,\ \xi \in \mathbb{R}^2.$$

Define a function $u \colon B(0,1) = \{y \in \mathbb{R}^2 \,:\, |y| < 1\} \to \mathbb{R}$ as

$$u(x) = |x|^{\alpha - 1} x_1, \quad \text{for } x = (x_1, x_2) \in B(0,1).$$

Then u is a weak solution (see the definition in Section 4.2) of equation (4.1) with the coefficients $\mathbb{A}(x)$ defined above (exercise).

4.1.2   Motivation: a variational problem

We start with two problems raised by Hilbert at the ICM in 1900.

- 20th problem: Has not every regular variational problem a solution, provided certain assumptions regarding the given boundary conditions are satisfied, and provided also if need be that the notions of a solution shall be suitably extended?

- 19th problem: Are the solutions of regular problems in the calculus of variations always necessarily analytic?

These problems are stated in a general way. We will consider the following specific variational problem to illustrate these problems and their solutions.

Let $\Omega \subset \mathbb{R}^n$ be a bounded smooth domain and $F \colon \mathbb{R}^n \to \mathbb{R}$ be a smooth function. We assume that it satisfies, for constants $0 < \lambda \leq \Lambda < \infty$,

$$\lambda |\xi|^2 \leq \langle D^2 F(\eta)\xi,\, \xi \rangle \leq \Lambda |\xi|^2, \quad \forall \xi \in \mathbb{R}^n,\ \eta \in \mathbb{R}^n. \tag{4.3}$$

Now we consider the functional

$$I(v) = \int_{\Omega} F(\nabla v)\, dx \tag{4.4}$$

among the admissible class

$$K' = \{v \,:\, v \in C^1(\bar{\Omega}) \text{ and } v = \phi \text{ on } \partial\Omega\},$$

where $\phi \in C^1(\bar{\Omega})$ is a given function. We say that $u \in K$ is a *minimizer* of the functional I among the class K if

$$I(u) \leq I(v), \quad \forall v \in K'.$$

Now the problems are the existence of minimizers (20th problem) and the regularity of minimizers (19th problem).

We cannot prove the existence of minimizers directly in the class K', due to the lack of compactness of the space $C^1(\bar{\Omega})$. We need to extend the space $C^1(\bar{\Omega})$ to a bigger space. The classical derivatives are extended to weak ones, and the classical solutions to weak ones, as suggested by Hilbert. A natural function space for this variational problem is the Sobolev space $W^{1,2}(\Omega)$. We offer a brief introduction to Sobolev spaces in Section 4.2. Now let

$$K = \{v \,:\, v \in W^{1,2}(\Omega) \text{ and } v - \phi \in W_0^{1,2}(\Omega)\}.$$

We can easily prove the existence of minimizers of the functional I in K by the direct method in the calculus of variations.

Theorem 4.3. *Suppose that $F \in C^\infty(\mathbb{R}^n)$ satisfies (4.3) and $\phi \in W^{1,2}(\Omega)$ is given. Then there is a unique $u_0 \in K$ such that*

$$I(u_0) = \inf_{u \in K} I(u).$$

Proof. Read Section 4.2 first. It is easy to show from (4.3) that (exercise)

$$c_1|\eta|^2 - c_2 \leq F(\eta) \leq c_3|\eta|^2 + c_4, \tag{4.5}$$

where $c_1 > 0$, $c_3 > 0$, c_2, c_4 are constants depending only on λ, Λ, $F(0)$, $\nabla F(0)$. Now let $m = \inf_{u \in K} I(u)$. It is easy to see that $-\infty < m < \infty$. Let $\{u_i\}_{i=1}^\infty \subset K$ be a minimizing sequence, that is, $I(u_i) \to m$ as $i \to \infty$. It is easy to prove that $\{u_i\}_{i=1}^\infty$ is a bounded sequence in $W^{1,2}(\Omega)$. Then, by the weak compactness theorem (Theorem 4.15), there is a subsequence, still denoted by itself, which converges weakly in $W^{1,2}(\Omega)$ to a function u_0. Since $u_i \in K$, we also have $u_0 \in K$. Now we claim that

$$\liminf_{i \to \infty} \int_\Omega F(\nabla u_i)\,dx \geq \int_\Omega F(\nabla u_0)\,dx,$$

which shows that $I(u_0) = m$ and hence u_0 is a minimizer of the functional I among the class K. Indeed, we have

$$F(\eta) \geq F(\eta_0) + \langle \nabla F(\eta_0), \eta - \eta_0 \rangle + \frac{\lambda}{2}|\eta - \eta_0|^2, \quad \forall \eta, \eta_0 \in \mathbb{R}^n. \tag{4.6}$$

The proof of (4.6) is as follows. We have that

$$F(\eta) - F(\eta_0) - \langle \nabla F(\eta_0), \eta - \eta_0 \rangle = \int_0^1 \frac{d}{dt} F(t\eta + (1-t)\eta_0)\, dt - \langle \nabla F(\eta_0), \eta - \eta_0 \rangle$$

$$= \int_0^1 \langle \nabla F(t\eta + (1-t)\eta_0) - \nabla F(\eta_0), \eta - \eta_0 \rangle\, dt,$$

and

$$\nabla F(t\eta + (1-t)\eta_0) - \nabla F(\eta_0) = \int_0^1 \frac{d}{ds} \nabla F(st\eta - st\eta_0 + \eta_0)\, ds$$

$$= t \int_0^1 D^2 F(st\eta - st\eta_0 + \eta_0)(\eta - \eta_0)\, ds.$$

Thus (4.6) follows from (4.3). Now, by (4.6), we have

$$\int_\Omega (F(\nabla u_i) - F(\nabla u_0))\, dx \geq \int_\Omega \langle \nabla F(\nabla u_0), \nabla u_i - \nabla u_0 \rangle\, dx.$$

The integral in the right-hand side goes to zero as i goes to infinity, since u_i converges to u_0 weakly in $W^{1,2}(\Omega)$. Thus the claim is true. It remains to prove uniqueness of the minimizer. Let u_0, $\bar{u}$ be minimizers and let $u = (u_0 + \bar{u})/2$. Note that $u \in K$. Then, by (4.6), we have that

$$F(\nabla u_0) \geq F(\nabla u) + \langle \nabla F(\nabla u), \nabla u_0 - \nabla u \rangle + \frac{\lambda}{2} |\nabla u - \nabla u_0|^2$$

and

$$F(\nabla \bar{u}) \geq F(\nabla u) + \langle \nabla F(\nabla u), \nabla \bar{u} - \nabla u \rangle + \frac{\lambda}{2} |\nabla u - \nabla \bar{u}|^2.$$

Adding these two inequalities together, we obtain that

$$F(\nabla u_0) + F(\nabla \bar{u}) \geq 2 F(\nabla u) + \frac{\lambda}{4} |\nabla u_0 - \nabla \bar{u}|^2.$$

Integrating both sides over Ω, we arrive at

$$2m \geq 2 \int_\Omega F(\nabla u)\, dx + \frac{\lambda}{4} \int_\Omega |\nabla u_0 - \nabla \bar{u}|^2\, dx \geq 2m + \frac{\lambda}{4} \int_\Omega |\nabla u_0 - \nabla \bar{u}|^2\, dx,$$

from which we deduce that $u_0 = \bar{u}$. This proves uniqueness. Hence, the theorem is proved.   □

The above theorem gives a positive solution to Hilbert's 20th problem. The good point about studying weak minimizers (solutions) is that it is easy to prove their existence, but the price we have to pay is the regularity. Now we turn to the

19th problem: regularity of minimizers. Our goal is to show that the minimizer u_0 is smooth. In order to do this, we first study the Euler–Lagrange equation corresponding to the functional I. It is easy to show that the minimizer u_0 that we obtained in Theorem 4.3 is a weak solution of the Euler–Lagrange equation. Note that, at this moment, we only know that u_0 is from the Sobolev space $W^{1,2}(\Omega)$.

Theorem 4.4. *Let u_0 be the minimizer given by Theorem 4.3. Then u_0 is a weak solution of the Euler–Lagrange equation*

$$\operatorname{div}(\nabla F(\nabla u_0)) = 0 \tag{4.7}$$

in Ω, that is,

$$\int_\Omega \langle \nabla F(\nabla u_0), \nabla \varphi \rangle \, dx = 0, \quad \forall \varphi \in C_0^\infty(\Omega). \tag{4.8}$$

Proof. Fix $\varphi \in C_0^\infty(\Omega)$. Define a function $g \colon \mathbb{R} \to \mathbb{R}$ as

$$g(t) = I(u_0 + t\varphi) = \int_\Omega F(\nabla u_0 + t\nabla\varphi) \, dx.$$

Then $g \in C^1(\mathbb{R})$ (prove it), and we have

$$g'(t) = \int_\Omega \langle \nabla F(\nabla u_0 + t\nabla\varphi), \nabla\varphi \rangle \, dx.$$

Since u_0 is a minimizer, the function g reaches its minimum at 0. Thus $g'(0) = 0$, which gives (4.8). $\qquad\square$

Next, we show that $u_0 \in W_{\mathrm{loc}}^{2,2}(\Omega)$ and that its weak derivatives are weak solutions of equation (4.1) with suitable coefficients $\mathbb{A}(x)$. In this way we build up a connection between this variational problem and equation (4.1).

Theorem 4.5. *Let u_0 be the minimizer given by Theorem 4.3. Then $u_0 \in W_{\mathrm{loc}}^{2,2}(\Omega)$ and $v_i = \partial_{x_i} u_0$, $i = 1, 2, \ldots, n$, is a weak solution of the equation*

$$\operatorname{div}(\mathbb{A}(x)\nabla v_i) = 0 \tag{4.9}$$

in Ω, where $\mathbb{A}(x) = D^2 F(\nabla u_0(x))$.

The formal proof involves difference quotients. Informally, we may just differentiate equation (4.7) with respect to x_i, $i = 1, 2, \ldots, n$, to obtain equation (4.9). The essential point of the proof is the Caccioppoli inequality, which will be discussed extensively in this note. Roughly speaking, the Caccioppoli inequality says that the L^2-norm of the derivatives of the solutions is controlled by that of the solutions. We refer to Section 4.2 for the notations in the proof.

Proof. Fix a cut-off function $\eta \in C_0^\infty(\Omega)$. Let Ω' be an open set such that $\mathrm{spt}(\eta) \subset \Omega' \Subset \Omega$. Let $\varphi = \Delta_i^{-h}((\Delta_i^h u_0)\eta^2)$, where $i = 1, 2, \ldots, n$ and h is so small that $0 < |h| < \mathrm{dist}(\Omega', \partial\Omega)/8$, and

$$\Delta_i^h v(x) = \frac{v(x + he_i) - v(x)}{h}$$

is the ith difference quotient of size h. Since $u_0 \in W^{1,2}(\Omega)$, we have $\varphi \in W_0^{1,2}(\Omega)$ (exercise). By Theorem 4.4, u_0 is a weak solution of equation (4.7). Thus,

$$0 = \int_\Omega \langle \nabla F(\nabla u_0), \nabla\varphi \rangle \, dx$$

$$= \int_\Omega \langle \nabla F(\nabla u_0), \Delta_i^{-h}(\nabla((\Delta_i^h u_0)\eta^2)) \rangle \, dx \qquad (4.10)$$

$$= -\int_\Omega \langle \Delta_i^h \nabla F(\nabla u_0), \nabla((\Delta_i^h u_0)\eta^2) \rangle \, dx,$$

where in the second equality we used the fact that ∇ and Δ_i^h are commutative, and in the last equality the so-called integration by parts for difference quotients. Now we write

$$\Delta_i^h \nabla F(\nabla u_0(x)) = \frac{\nabla F(\nabla u_0(x + he_i)) - \nabla F(\nabla u_0(x))}{h}$$

$$= \frac{1}{h} \int_0^1 \frac{d}{dt} \nabla F(t\nabla u_0(x + he_i) + (1 - t)\nabla u_0(x)) \, dt$$

$$= \int_0^1 D^2 F(t\nabla u_0(x + he_i) + (1 - t)\nabla u_0(x)) \, dt \, \Delta_i^h \nabla u_0(x)$$

$$= \mathbb{B}(x)\Delta_i^h \nabla u_0(x),$$

and

$$\nabla((\Delta_i^h u_0)\eta^2) = \eta^2 \Delta_i^h \nabla u_0 + 2\eta \nabla\eta \Delta_i^h u_0.$$

Thus (4.10) becomes

$$\int_\Omega \langle \mathbb{B}(x)\Delta_h^i \nabla u_0, \Delta_i^h \nabla u_0 \rangle \eta^2 \, dx = -2 \int_\Omega \langle \mathbb{B}(x)\Delta_i^h \nabla u_0, \nabla\eta \rangle \eta \Delta_i^h u_0 \, dx.$$

Note that the matrix $\mathbb{B}(x)$ also satisfies (4.2). By the Cauchy–Schwarz inequality and Hölder's inequality, we can easily deduce the following Caccioppoli type inequality:

$$\int_\Omega |\Delta_i^h \nabla u_0|^2 \eta^2 \, dx \leq \frac{4\Lambda}{\lambda} \int_\Omega |\Delta_i^h u_0|^2 |\nabla\eta|^2 \, dx,$$

which, together with Theorem 4.17, proves the theorem. $\square$

One question here: Is it possible to repeat the above argument to prove that $u_0 \in W^{3,2}(\Omega)$? The answer is no. One may informally differentiate equation (4.9) and see if we can obtain a Caccioppoli type inequality to control the L^2-norm of the third-order derivatives of u_0.

Finally, our main goal is to prove that the weak solutions of equation (4.1) are Hölder continuous.

Theorem 4.6. *Let $u \in W^{1,2}(\Omega)$ be a weak solution of equation (4.1). Then $u \in C^{0,\alpha}(\Omega)$, where $0 < \alpha = \alpha(n, \lambda, \Lambda) \le 1$.*

In the planar case, the study goes back to the work of Morrey [11, 12]; see [10], [17] and [21] for the study of the best Hölder continuity exponent. In higher dimensions ($\mathbb{R}^n$, $n \geqslant 3$), Hölder continuity of solutions was settled in the late 1950's by De Giorgi [1] and Nash [15]. Hölder continuity also follows from the Harnack inequality, due to Moser [13, 14].

Now we go back to our variational problem. Combining Theorem 4.4 and Theorem 4.6, we obtain that $u_0 \in C^{1,\alpha}(\Omega)$ for some $\alpha > 0$. Then we can show that actually $u_0 \in C^{\infty}(\Omega)$, by the Schauder estimates.

The above is the line to deal with Hilbert's 20th and 19th problems. As we can see, the essential and difficult point is the De Giorgi–Nash–Moser theory: the Hölder continuity of solutions of equation (4.1) in Theorem 4.6. In these notes, we first explain Moser's method and then De Giorgi's method. The reader is expected to compare these two methods. While these two methods are further exploited and applied to many other problems, the argument of Nash is rather difficult to penetrate and consequently his work has not been extensively used in the literature. We refer to [6] for the applications of Nash's ideas.

One comment: To the author's knowledge, De Giorgi's, Nash's and Moser's methods are the only existing approaches to prove Hölder continuity of weak solutions of equation (4.1). I do not know any other way to prove Theorem 4.6.

4.2 Sobolev spaces

4.2.1 A brief introduction to Sobolev spaces

We refer to [4], [7] and [9] for the proofs of theorems in this subsection. Let Ω be an open set in $\mathbb{R}^n$.

Hölder space

Definition 4.7. We say that a function u belongs to $C^{0,\alpha}(\Omega)$ if, for every open set $\Omega' \Subset \Omega$,

$$||u||_{C^{0,\alpha}(\bar{\Omega}')} = \sup_{x \in \Omega'} |u(x)| + \sup_{x,y \in \Omega', x \neq y} \frac{|u(x) - u(y)|}{|x - y|^\alpha} < \infty.$$

Weak derivatives

Let $C_0^\infty(\Omega)$ denote the space of infinitely differentiable functions with compact support in Ω.

Definition 4.8. Suppose that $u, v \in L^1_{\text{loc}}(\Omega)$. We say that v is the *ith weak partial derivative* of u, written

$$\partial_{x_i} u = v,$$

provided that

$$\int_\Omega u \partial_{x_i} \phi \, dx = -\int_\Omega v\phi \, dx$$

for all test functions $\phi \in C_0^\infty(\Omega)$.

A weak partial derivative of a function u, if it exists, is uniquely defined up to a set of measure zero. We denote by $\nabla u = (\partial_{x_1} u, \partial_{x_2} u, \ldots, \partial_{x_n} u)$ the weak gradient of u.

Definition of Sobolev spaces

Definition 4.9. The *Sobolev space*

$$W^{1,p}(\Omega), \quad 1 \le p \le \infty,$$

consists of all locally integrable functions $u\colon \Omega \to \mathbb{R}$ such that $u \in L^p(\Omega)$ and the weak derivatives $\partial_{x_i} u$ are in $L^p(\Omega)$ for all $i = 1, 2, \ldots, n$. We define its *norm* to be

$$\|u\|_{W^{1,p}(\Omega)} = \begin{cases} \left(\displaystyle\int_\Omega |u|^p \, dx + \sum_{i=1}^n \int_\Omega |\partial_{x_i} u|^p \, dx \right)^{1/p} & \text{if } 1 \le p < \infty; \\ \displaystyle\sup_\Omega |u| + \sum_{i=1}^n \sup_\Omega |\partial_{x_i} u| & \text{if } p = \infty. \end{cases}$$

We denote by $W_0^{1,p}(\Omega)$ the closure of $C_0^\infty(\Omega)$ in $W^{1,p}(\Omega)$ with respect to the norm defined above. Thus $u \in W_0^{1,p}(\Omega)$ if and only if there exist functions $u_k \in C_0^\infty(\Omega)$ such that

$$u_k \to u \quad \text{in } W^{1,p}(\Omega),$$

that is,

$$\lim_{k\to\infty} \|u_k - u\|_{W^{1,p}(\Omega)} = 0.$$

Exercise 4.10. Prove that $W^{1,p}(\Omega)$ and $W_0^{1,p}(\Omega)$ are Banach spaces.

Inequalities

Theorem 4.11 (Gagliardo–Nirenberg–Sobolev inequality). *Assume that $1 \leq p < n$. There is a constant $c = c(n, p) > 0$ such that*

$$\left(\int_{\Omega} |u|^{p^*} \, dx \right)^{1/p^*} \leq c \left(\int_{\Omega} |\nabla u|^p \, dx \right)^{1/p}$$

for all $u \in W_0^{1,p}(\Omega)$, where $p^* = np/(n - p)$.

We also have the following version of Sobolev's inequality.

Theorem 4.12. *Assume that $1 \leq p < n$. Suppose that $u \in W^{1,p}(B(y, r))$ and $|\{x \in B(y, r) : u(x) = 0\}| \geq \delta |B(y, r)|$ for some $\delta > 0$. Then*

$$\left(\int_{B(y,r)} |u|^{p^*} \, dx \right)^{1/p^*} \leq c \left(\int_{B(y,r)} |\nabla u|^p \, dx \right)^{1/p},$$

where $c = c(n, p, \delta) > 0$.

The following theorem is an easy consequence of Theorem 4.12.

Theorem 4.13. *Suppose that $u \in W^{1,1}(B(y, r))$ and $|\{x \in B(y, r) : u(x) \leq k\}| \geq \delta |B(y, r)|$ for some $\delta > 0$ and $k \in \mathbb{R}$. Then*

$$(l - k)|\{x \in B(y, r) : u(x) > l\}|^{1 - \frac{1}{n}} \leq c \int_{k < u < l} |\nabla u| \, dx$$

for any $l > k$, where $c = c(n, \delta) > 0$.

Theorem 4.14 (Poincaré inequality). *Assume that $1 \leq p < \infty$. Then there is a constant $c = c(n, p) > 0$ such that*

$$\int_{B(y,r)} |u - u_{B(y,r)}|^p \, dx \leq c r^p \int_{B(y,r)} |\nabla u|^p \, dx$$

for every $u \in W^{1,p}(\Omega)$ and every ball $B(y, r) \subset \Omega$, where $u_{B(y,r)} = \fint_{B(y,r)} u \, dx$ is the average of u over the ball $B(y, r)$.

Weak compactness theorem

Theorem 4.15 (Weak compactness theorem). *Assume $1 < p < \infty$. Suppose that $\{u_k\}_{k=1}^{\infty}$ is a bounded sequence in $W^{1,p}(\Omega)$. Then there are a function $u \in W^{1,p}(\Omega)$ and a subsequence of $\{u_k\}$, still denoted by itself, such that u_k converges weakly in $W^{1,p}(\Omega)$ to u, that is, we have that*

$$\lim_{k \to \infty} \int_{\Omega} u_k \phi \, dx = \int_{\Omega} u \phi \, dx, \quad \forall \phi \in L^{\frac{p}{p-1}}(\Omega),$$

and

$$\lim_{k \to \infty} \int_{\Omega} \partial_{x_i} u_k \phi \, dx = \int_{\Omega} \partial_{x_i} u \phi \, dx, \quad \forall \phi \in L^{\frac{p}{p-1}}(\Omega), \; i = 1, 2, \ldots, n.$$

Difference quotients

Let $v \colon \Omega \to \mathbb{R}$ be a locally integrable function, and $\Omega' \Subset \Omega$.

Definition 4.16. The *ith difference quotient of size h* is

$$\Delta_i^h v(x) = \frac{v(x + he_i) - v(x)}{h}, \quad i = 1, 2, \dots, n,$$

for $x \in \Omega'$, $h \in \mathbb{R}$, $0 < |h| < \mathrm{dist}(\Omega', \partial\Omega)$. We write $\Delta^h v = (\Delta_1^h v, \Delta_2^h v, \dots, \Delta_n^h v)$.

Theorem 4.17 (Difference quotients and weak derivatives).

(i) *Assume that $1 \le p < \infty$ and $v \in W^{1,p}(\Omega)$. Then, for every $\Omega' \Subset \Omega$,*

$$\int_{\Omega'} |\Delta^h v|^p \, dx \le \int_{\Omega} |\nabla v|^p$$

whenever $0 < |h| < \mathrm{dist}(\Omega', \partial\Omega)/2$.

(ii) *Assume that $1 < p < \infty$ and $u \in L^p(\Omega')$. Suppose that there is a constant c such that*

$$\int_{\Omega'} |\Delta^h v|^p \, dx \le c$$

whenever $0 < |h| < \mathrm{dist}(\Omega', \partial\Omega)/2$. Then

$$v \in W^{1,p}(\Omega') \quad \text{and} \quad \int_{\Omega'} |\nabla v|^p \, dx \le c.$$

4.2.2   Definition of weak solutions

Definition 4.18. We say that a function $u \in W^{1,2}(\Omega)$ is a *weak solution* of equation (4.1) if

$$\int_{\Omega} \langle \mathbb{A}(x)\nabla u, \nabla\varphi \rangle \, dx = \int_{\Omega} \sum_{i,j=1}^{n} a_{ij}(x) \, \partial_{x_j} u \, \partial_{x_i}\varphi \, dx = 0 \qquad (4.11)$$

for every $\varphi \in C_0^\infty(\Omega)$. We say that a function $u \in W^{1,2}(\Omega)$ is a *weak subsolution* of equation (4.1) if

$$\int_{\Omega} \langle \mathbb{A}(x)\nabla u, \nabla\varphi \rangle \, dx \le 0$$

for every nonnegative $\varphi \in C_0^\infty(\Omega)$. We define *weak supersolutions* similarly.

Remark 4.19.

(i) If u is a classical solution of equation (4.1), then it is a weak solution.

(ii) In the definition, we require that the weak solutions are from the Sobolev space $W^{1,2}(\Omega)$. This is the natural Sobolev space for the definition of weak solutions of equation (4.1). Actually, the formula (4.11) makes sense if we only require that $u \in W^{1,p}(\Omega)$ for some $p \ge 1$. We call this kind of solutions *very weak solutions*. See [8] for the study of very weak solutions.

(iii) Since in the definition we require that $u \in W^{1,2}(\Omega)$, it is easy to prove by an approximation argument that (4.11) holds for all $\varphi \in W_0^{1,2}(\Omega)$ (exercise).

4.3 Moser's iteration

In this section we prove Theorem 4.6. In what follows, $u \in W^{1,2}(\Omega)$ is a weak solution of the equation

$$\operatorname{div}(\mathbb{A}(x)\nabla u(x)) = 0, \tag{4.12}$$

where $\mathbb{A}(x)$ is a symmetric matrix satisfying, for $0 < \lambda \le \Lambda < \infty$,

$$\lambda|\xi|^2 \le \langle \mathbb{A}(x)\xi, \xi \rangle \le \Lambda|\xi|^2, \quad \forall \xi = (\xi_1, \dots, \xi_n) \in \mathbb{R}^n, \text{ a.e. } x \in \Omega. \tag{4.13}$$

4.3.1 Harnack's inequality

We first prove the following Harnack inequality.

Theorem 4.20. *Let $u \in W^{1,2}(\Omega)$, $u \ge 0$ in Ω, be a weak solution of equation (4.12). Then there is a constant $c = c(n, \lambda, \Lambda) > 0$ such that for every ball $B(y,r) \subset \Omega$ we have*

$$\sup_{B(y,r/2)} u \le c \inf_{B(y,r/2)} u.$$

The proof of Harnack's inequality is divided into two parts in Subsection 4.3.2 and Subsection 4.3.3. The Hölder continuity of solutions is an easy consequence of Harnack's inequality. We leave the proof as an exercise. We use the notation $\operatorname{osc}_{B(y,t)} u = \sup_{B(y,t)} u - \inf_{B(y,t)} u$.

Theorem 4.21. *Let $u \in W^{1,2}(\Omega)$ be a weak solution of equation (4.12). Then there is $\alpha = \alpha(n, \lambda, \Lambda)$, $0 < \alpha \le 1$, such that $u \in C^{0,\alpha}(\Omega)$. Moreover, for every ball $B(y, R) \subset \Omega$ and all $0 < r \le R < \infty$, we have*

$$\operatorname{osc}_{B(y,r)} u \le 2^\alpha \left(\frac{r}{R}\right)^\alpha \operatorname{osc}_{B(y,R)} u.$$

4.3.2 Weak Harnack's inequality: sup

We next prove the local boundedness of weak solutions.

Theorem 4.22. *Let $u \in W^{1,2}(\Omega)$ be a weak solution of equation (4.12). Then $u \in L^\infty_{\text{loc}}(\Omega)$. Moreover, for every ball $B(y,r) \subset \Omega$ we have*

$$\sup_{B(y,r/2)} |u| \le c \left(\fint_{B(y,r)} |u|^2 \, dx\right)^{\frac{1}{2}}, \tag{4.14}$$

where $c = c(n, \lambda, \Lambda) > 0$.

The essential ingredients of the proof are a Caccioppoli type inequality and the Sobolev inequality. An iteration argument is involved. The starting point is the following Caccioppoli inequality.

Lemma 4.23. *Let* $u \in W^{1,2}(\Omega)$ *be a weak solution of equation* (4.12). *Then for any* $\alpha \geq 0$ *and any* $\eta \in C_0^\infty(\Omega)$ *we have*

$$\int_\Omega |u|^\alpha |\nabla u|^2 \eta^2 \, dx \leq c \int_\Omega |u|^{\alpha+2} |\nabla \eta|^2 \, dx, \qquad (4.15)$$

where $c = c(\lambda, \Lambda) > 0$, *provided that* $u \in L_{\mathrm{loc}}^{\alpha+2}(\Omega)$.

Proof. Fix $\eta \in C_0^\infty(\Omega)$. Let $t \geq 0$ and define $v = (u - t)^+ = \max(u - t, 0)$. Using $\varphi = v\eta^2 \in W_0^{1,2}(\Omega)$ (prove it) as a test function in equation (4.12), we obtain that

$$\begin{aligned}
0 &= \int_\Omega \langle \mathbb{A}(x)\nabla u, \nabla \varphi \rangle \, dx \\
&= \int_\Omega \langle \mathbb{A}(x)\nabla u, \nabla(u - t)^+ \rangle \eta^2 \, dx + 2 \int_\Omega \langle \mathbb{A}(x)\nabla u, \nabla \eta \rangle (u - t)^+ \eta \, dx.
\end{aligned} \qquad (4.16)$$

We use the Cauchy–Schwarz inequality

$$|\langle \mathbb{A}(x)\nabla u, \nabla \eta \rangle| \leq \langle \mathbb{A}(x)\nabla u, \nabla u \rangle^{\frac{1}{2}} \langle \mathbb{A}(x)\nabla \eta, \nabla \eta \rangle^{\frac{1}{2}}$$

and Hölder's inequality to estimate the last integral. Then (4.16) gives us

$$\int_{u>t} \langle \mathbb{A}(x)\nabla u, \nabla u \rangle \eta^2 \, dx \leq 4 \int_{u>t} \langle \mathbb{A}(x)\nabla \eta, \nabla \eta \rangle |(u - t)^+|^2 \, dx,$$

which, together with (4.13), yields

$$\int_{u>t} |\nabla u|^2 \eta^2 \, dx \leq \frac{4\Lambda}{\lambda} \int_{u>t} |(u - t)^+|^2 |\nabla \eta|^2 \, dx \leq \frac{4\Lambda}{\lambda} \int_{u>t} |u^+|^2 |\nabla \eta|^2 \, dx. \quad (4.17)$$

Now the above inequality holds for all $t \geq 0$. We multiply both sides by $\alpha t^{\alpha-1}$ and integrate with respect to t over $(0, \infty)$. A direct calculation gives

$$\int_\Omega |u|^\alpha |\nabla u^+|^2 \eta^2 \, dx \leq \frac{4\Lambda}{\lambda} \int_\Omega |u^+|^{\alpha+2} |\nabla \eta|^2 \, dx.$$

Similarly, the above inequality is also true for u^-. Then we obtain (4.15) with $c = 4\Lambda/\lambda$. This proves the lemma.   □

Now, by the Sobolev inequality, we obtain the following reverse inequality.

Lemma 4.24. *Let* $u \in W^{1,2}(\Omega)$ *be a weak solution of equation* (4.12). *Then* $u \in L_{\mathrm{loc}}^{(\alpha+2)\chi}(\Omega)$ *if* $u \in L_{\mathrm{loc}}^{\alpha+2}(\Omega)$ *for any* $\alpha \geq 0$, *where* $\chi = n/(n - 2)$ *when* $n \geq 3$. *Moreover, for any* $\eta \in C_0^\infty(\Omega)$,

$$\left(\int_\Omega |u|^{(\alpha+2)\chi} \eta^{2\chi} \, dx \right)^{1/\chi} \leq c(\alpha + 2)^2 \int_\Omega |u|^{\alpha+2} |\nabla \eta|^2 \, dx, \qquad (4.18)$$

where $c = c(n, \lambda, \Lambda) > 0$.

Proof. Let $v = |u|^{\alpha/2} u\eta$. Then

$$\nabla v = \left(\frac{\alpha}{2} + 1\right) |u|^{\alpha/2} \eta \nabla u + |u|^{\alpha/2} u \nabla \eta.$$

Thus (4.15) in Lemma 4.23 gives us

$$\int_\Omega |\nabla v|^2 \, dx \le c(\alpha + 2)^2 \int_\Omega |u|^{\alpha+2} |\nabla \eta|^2 \, dx, \qquad (4.19)$$

where $c = c(\lambda, \Lambda) > 0$. Now we use the Sobolev inequality in Theorem 4.11 (when $n \ge 3$):

$$\left(\int_\Omega |v|^{2\chi} \, dx\right)^{1/\chi} \le c(n) \int_\Omega |\nabla v|^2 \, dx. \qquad (4.20)$$

Combining (4.19) and (4.20) yields (4.18). This finishes the proof. $\square$

The following corollary is an easy consequence of Lemma 4.24.

Corollary 4.25. *Let $u \in W^{1,2}(\Omega)$ be a weak solution of equation (4.12). Then $u \in L^q_{\mathrm{loc}}(\Omega)$ for every $q \ge 1$. Moreover, for every $\alpha \ge 0$, for every ball $B(y, r) \subset \Omega$ and for every $0 < r' < r$, we have the following reverse type inequality:*

$$\left(\fint_{B(y,r')} |u|^{(\alpha+2)\chi} \, dx\right)^{\frac{1}{(\alpha+2)\chi}} \le \frac{c^{\frac{1}{\alpha+2}}(\alpha+2)^{\frac{2}{\alpha+2}}}{(r-r')^{\frac{2}{\alpha+2}}} \left(\fint_{B(y,r)} |u|^{\alpha+2} \, dx\right)^{\frac{1}{\alpha+2}}, \quad (4.21)$$

where $c = c(n, \lambda, \Lambda) > 0$.

Now we iterate (4.21) to prove Theorem 4.22.

Proof of Theorem 4.22. Fix a ball $B(y, r) \subset \Omega$. Define $\alpha_0 = 0$, $\alpha_i = 2\chi^i - 2$, $i = 1, 2, \ldots$. Let $r_0 = r$ and

$$r_i = \frac{r}{2} + \frac{r}{2^{i+1}}, \quad i = 1, 2, \ldots.$$

Apply (4.21) with $r = r_i$, $r' = r_{i+1}$ and $\alpha = \alpha_i$ for $i = 0, 1, \ldots$. We obtain that

$$M_{i+1} \le c^{1/\beta_i} \beta_i^{2/\beta_i} \left(\frac{r}{2^{i+2}}\right)^{-2/\beta_i} M_i, \qquad (4.22)$$

where $\beta_i = 2\chi^i$ and

$$M_i = \left(\fint_{B(y,r_i)} |u|^{\beta_i} \, dx\right)^{1/\beta_i}.$$

An iteration of (4.22) gives us

$$M_{i+1} \le c_i M_0,$$

from which we obtain (4.14) by letting $i \to \infty$. We leave the details as an exercise. This concludes the proof of Theorem 4.22. $\square$

Slightly modifying the above argument, we can prove the following version of Theorem 4.22 for the weak subsolutions (see Definition 4.18) when $n \geq 3$. Write down the details of the proof as an exercise.

Theorem 4.26. Let $u \in W^{1,2}(\Omega)$, $u \geq 0$ in Ω, be a weak subsolution of equation (4.12). Then $u \in L^{\infty}_{\text{loc}}(\Omega)$. Moreover, for every ball $B(y,r) \subset \Omega$ and $0 < \sigma < 1$, we have

$$\sup_{B(y,\sigma r)} u \leq \frac{c}{(1-\sigma)^{n/2}} \left(\fint_{B(y,r)} u^2 \, dx \right)^{\frac{1}{2}}, \qquad (4.23)$$

where $c = c(n, \lambda, \Lambda) > 0$.

Finally, by another iteration argument, we prove the following boundedness estimate for subsolutions.

Theorem 4.27. Let $u \in W^{1,2}(\Omega)$, $u \geq 0$ in Ω, be a weak subsolution of equation (4.12). Then $u \in L^{\infty}_{\text{loc}}(\Omega)$. Moreover, for every ball $B(y,r) \subset \Omega$ and $0 < \sigma < 1$, $0 < q \leq 2$, we have

$$\sup_{B(y,\sigma r)} u \leq \frac{c}{(1-\sigma)^{n/q}} \left(\fint_{B(y,r)} u^q \, dx \right)^{1/q}, \qquad (4.24)$$

where $c = c(q, n, \lambda, \Lambda) > 0$.

Proof. Fix $B(y,r) \subset \Omega$ and $0 < \sigma < 1$. Let $\sigma_0 = \sigma$ and

$$\sigma_i = 1 - \frac{1-\sigma}{2^i}, \quad i = 1, 2, \dots.$$

By Theorem 4.26, (4.23) with $r = \sigma_{i+1} r$ and $\sigma = \sigma_i / \sigma_{i+1}$ gives us

$$\sup_{B(y,\sigma_i r)} u \leq \frac{c}{\left(1 - \frac{\sigma_i}{\sigma_{i+1}}\right)^{n/2}} \left(\fint_{B(y,\sigma_{i+1} r)} u^2 \, dx \right)^{\frac{1}{2}}$$

$$\leq \frac{c}{\left(1 - \frac{\sigma_i}{\sigma_{i+1}}\right)^{n/2}} \left(\fint_{B(y,\sigma_{i+1} r)} u^q \, dx \right)^{\frac{1}{2}} \left(\sup_{B(y,\sigma_{i+1})} u \right)^{\frac{2-q}{2}},$$

that is,

$$M_i \leq \frac{c}{\left(1 - \frac{\sigma_i}{\sigma_{i+1}}\right)^{n/2}} \left(\fint_{B(y,r)} u^q \, dx \right)^{\frac{1}{2}} M_{i+1}^{\frac{2-q}{2}}, \qquad (4.25)$$

where $M_i = \sup_{B(y,\sigma_i r)} u$. An iteration of (4.25) gives (4.14). We leave the details as an exercise. Thus the proof is complete. $\qquad \square$

Exercise 4.28. Figure out a version of Theorem 4.27 in the case $n = 2$ and write down a proof.

4.3.3 Weak Harnack's inequality: inf

In this subsection, we prove the following result.

Theorem 4.29. *Let $u \in W^{1,2}(\Omega)$, $u \geq 0$ in Ω, be a weak solution of equation (4.12). Then there are $q = q(n, \lambda, \Lambda) > 0$ and $c = c(n, \lambda, \Lambda) > 0$ such that, for every ball $B(y, 2r) \subset \Omega$, we have*

$$\inf_{B(y,r/2)} u \geq c \left(\fint_{B(y,r)} u^q \, dx \right)^{1/q}. \qquad (4.26)$$

By replacing u by $u + \varepsilon$ for $\varepsilon > 0$, we may assume that $u \geq \varepsilon$ in Ω. The essential point of the proof of Theorem 4.29 is that $\log u$ is a function of bounded mean oscillation (BMO). First, Theorem 4.27 yields the following estimate.

Lemma 4.30. *Let $u \in W^{1,2}(\Omega)$ be a weak solution of equation (4.12). Suppose that $u \geq \varepsilon$ in Ω for some $\varepsilon > 0$. Then for any $q > 0$ there is $c = c(q, n, \lambda, \Lambda) > 0$ such that, for every ball $B(y, r) \subset \Omega$, we have*

$$\inf_{B(y,r/2)} u \geq c \left(\fint_{B(y,r)} u^{-q} \, dx \right)^{-1/q}. \qquad (4.27)$$

Proof. We claim that $v = 1/u$ is a subsolution of equation (4.12). Indeed, first, it is easy to show that $1/u \in W^{1,2}_{\text{loc}}(\Omega)$. Second, for any $\eta \in C_0^\infty(\Omega)$, $\eta \geq 0$, let $\varphi = \eta/u^2$. We use φ as a test function in equation (4.12) to obtain that

$$0 = \int_\Omega \langle \mathbb{A}(x)\nabla u, \nabla \varphi \rangle \, dx$$

$$= \int_\Omega \langle \mathbb{A}(x)\nabla u, \nabla \eta \rangle \frac{1}{u^2} \, dx - 2 \int_\Omega \langle \mathbb{A}(x)\nabla u, \nabla u \rangle \frac{\eta}{u^3} \, dx.$$

The last integral is nonnegative. Thus we have

$$\int_\Omega \langle \mathbb{A}(x)\nabla v, \nabla \eta \rangle \, dx = - \int_\Omega \langle \mathbb{A}(x)\nabla u, \nabla \eta \rangle \frac{1}{u^2} \, dx \leq 0,$$

which shows that $1/u$ is a subsolution. Therefore, the lemma follows from Theorem 4.27. $\qquad \square$

Second, we show that $\log u$ is of BMO.

Lemma 4.31. *Let $u \in W^{1,2}(\Omega)$ be a weak solution of equation (4.12). Suppose that $u \geq \varepsilon$ in Ω for some $\varepsilon > 0$. Then for every ball $B(y, 2r) \subset \Omega$ we have*

$$\int_{B(y,r)} |\nabla v|^2 \, dx \leq c r^{n-2}, \qquad (4.28)$$

where $v = \log u$ and $c = c(n, \lambda, \Lambda) > 0$.

Proof. Fix $\eta \in C_0^\infty(\omega)$ and let $\varphi = \eta^2/u$. We use φ as a test function in equation (4.12) to obtain that

$$0 = \int_\Omega \langle \mathbb{A}(x)\nabla u, \nabla \varphi \rangle \, dx$$

$$= - \int_\Omega \langle \mathbb{A}(x)\nabla u, \nabla u \rangle \frac{\eta^2}{u^2} \, dx + 2 \int_\Omega \langle \mathbb{A}(x)\nabla u, \nabla \eta \rangle \frac{\eta}{u} \, dx,$$

from which we deduce that

$$\int_\Omega |\nabla v|^2 \eta^2 \leq \frac{4\Lambda}{\lambda} \int_\Omega |\nabla \eta|^2 \, dx. \tag{4.29}$$

Then (4.28) follows by choosing $\eta \in C_0^\infty(B(y,2r))$ such that $\eta = 1$ on $B(y,r)$ and $|\nabla \eta| \leq 2/r$ in $B(y,2r)$. $\square$

Definition 4.32. A function $v \in L^1_{\mathrm{loc}}(\Omega)$ is said to be of *bounded mean oscillation*, denoted by $v \in BMO(\Omega)$, if

$$[v]_{BMO} = \sup \fint_{B(y,r)} |v - v_{B(y,r)}| \, dx < \infty, \tag{4.30}$$

where $v_{B(y,r)} = \fint_{B(y,r)} v \, dx$ is the average of v over the ball $B(y,r)$ and the supremum in (4.30) is taken for all balls $B(y,r)$ such that $B(y,2r) \subset \Omega$.

A fundamental property of BMO functions is exponential integrability.

Lemma 4.33 (John–Nirenberg lemma). *Suppose that $v \in BMO(\Omega)$. Then there are positive constants c_1 and c_2, depending only on n and $[v]_{BMO}$, such that for every $B(y,2r) \subset \Omega$ we have*

$$\fint_{B(y,r)} \exp(c_1|v - v_{B(y,r)}|) \, dx \leq c_2. \tag{4.31}$$

Finally, we prove Theorem 4.29.

Proof of Theorem 4.29. By Lemma 4.31 and the Poincaré inequality, we have, for every ball $B(y,2r) \subset \Omega$,

$$\fint_{B(y,r)} |v - v_{B(y,r)}|^2 \, dx \leq c(n)r^2 \fint_{B(y,r)} |\nabla v|^2 \, dx \leq c(n,\lambda,\Lambda).$$

Thus $u \in BMO(\Omega)$. Then the John–Nirenberg lemma yields

$$\fint_{B(y,r)} \exp(c_1|v - v_{B(y,r)}|) \, dx \leq c_2$$

for $c_1 = c_1(n, \lambda, \Lambda) > 0$ and $c_2 = c_2(n, \lambda, \Lambda) > 0$. Then we have

$$\fint_{B(y,r)} u^{c_1} \, dx \fint_{B(y,r)} u^{-c_1} \, dx$$

$$= \fint_{B(y,r)} \exp(c_1(v - v_{B(y,r)})) \, dx \fint_{B(y,r)} \exp(c_1(v_{B(y,r)} - v)) \, dx$$

$$\leq \left(\fint_{B(y,r)} \exp(c_1 |v - v_{B(y,r)}|) \, dx \right)^2 \leq (c_2)^2,$$

which, together with Lemma 4.30, proves (4.26) with $q = c_1$. This finishes the proof of Theorem 4.29. $\qquad\square$

4.4   De Giorgi's method

4.4.1   De Giorgi's class of functions

In his fundamental work on linear elliptic equations, De Giorgi [1] established local boundedness and Hölder continuity for functions satisfying certain integral inequalities, known as the De Giorgi class of functions.

Let Ω be an open set in $\mathbb{R}^n$ and γ be a constant. The De Giorgi class $DG^+(\Omega, \gamma)$ consists of functions $u \in W^{1,2}(\Omega)$ which satisfy, for every ball $B(y, r) \subset \Omega$, every $0 < r' < r$, and every $k \in \mathbb{R}$, the following Caccioppoli type inequality:

$$\fint_{B(y,r')} |\nabla(u - k)^+|^2 \, dx \leq \frac{\gamma}{(r - r')^2} \fint_{B(y,r)} |(u - k)^+|^2 \, dx, \qquad (4.32)$$

where $(u - k)^+ = \max(u - k, 0)$. Similarly, we may define the class $DG^-(\Omega, \gamma)$ by replacing $(u - k)^+$ by $(u - k)^- = \min(u - k, 0)$. Thus $u \in DG^+(\Omega, \gamma)$ if and only if $-u \in DG^-(\Omega, \gamma)$. We denote $DG(\Omega, \gamma) = DG^+(\Omega, \gamma) \cap DG^-(\Omega, \gamma)$.

All weak solutions of equation (4.12) are in the De Giorgi class. We already proved the following lemma; see the proof of Lemma 4.23.

Lemma 4.34. *Let $u \in W^{1,2}(\Omega)$ be a weak subsolution of equation (4.12). Then $u \in DG^+(\Omega, \gamma)$ for some $\gamma = \gamma(\lambda, \Lambda) > 0$.*

4.4.2   Boundedness of functions in $DG(\Omega, \gamma)$

Theorem 4.35. *Let $\gamma > 0$ be a constant and $u \in DG^+(\Omega, \gamma)$. Then $u \in L^\infty_{\text{loc}}(\Omega)$. Moreover, for every ball $B(y, r) \subset \Omega$, we have*

$$\sup_{B(y, r/2)} u^+ \leq c \left(\fint_{B(y,r)} |u^+|^2 \, dx \right)^{\frac{1}{2}}, \qquad (4.33)$$

where $c = c(n, \gamma) > 0$.

Proof. Fix $B(y,r) \subset \Omega$. Let $M > 0$ be a number to be chosen later. We set

$$k_i = M - \frac{M}{2^i}, \quad i = 0, 1, 2, \dots$$

and consider the sequence of radii

$$r_i = \frac{r}{2} + \frac{r}{2^{i+1}}, \quad \bar{r}_i = \frac{1}{2}(r_i + r_{i+1}) = \frac{r}{2} + \frac{3}{4} \cdot \frac{r}{2^{i+1}}, \quad i = 0, 1, 2, \dots.$$

Let $\eta_i \in C_0^\infty(B(y, \bar{r}_i))$ be a cut-off function such that $\eta_i = 1$ in $B(y, r_{i+1})$ and $|\nabla \eta_i| \le 2^{i+8}/r$. We only prove the case $n \ge 3$. By Hölder's inequality, we have

$$\int_{B(y, r_{i+1})} |(u - k_{i+1})^+|^2 \, dx \le \int_{B(y, \bar{r}_i)} |(u - k_{i+1})^+ \eta_i|^2 \, dx$$

$$\le \left(\int_{B(y, \bar{r}_i)} |(u - k_{i+1})^+ \eta_i|^{\frac{2n}{n-2}} \, dx \right)^{\frac{n-2}{n}} |A_i|^{\frac{2}{n}}, \tag{4.34}$$

where $A_i = \{x \in B(y, \bar{r}_i) : u(x) > k_{i+1}\}$. We continue to estimate the first integral on the right-hand side of (4.34) by means of the Sobolev inequality.

$$I = \left(\int_{B(y, \bar{r}_i)} |(u - k_{i+1})^+ \eta_i|^{\frac{2n}{n-2}} \, dx \right)^{\frac{n-2}{n}}$$

$$\le c(n) \int_{B(y, \bar{r}_i)} |\nabla((u - k_{i+1})^+ \eta_i)|^2 \, dx$$

$$\le c \int_{B(y, \bar{r}_i)} |\nabla(u - k_i)^+|^2 \, dx + c \int_{B(y, \bar{r}_i)} |(u - k_{i+1})^+|^2 \, |\nabla \eta_i|^2 \, dx.$$

Since $u \in DG^+(\Omega, \gamma)$, we have

$$\int_{B(y, \bar{r}_i)} |\nabla(u - k_i)^+|^2 \, dx \le \frac{\gamma}{(r_i - \bar{r}_i)^2} \int_{B(y, r_i)} |(u - k_i)^+|^2 \, dx.$$

Thus, we get

$$I \le \frac{c 2^{2i}}{r^2} \int_{B(y, r_i)} |(u - k_i)^+|^2 \, dx, \tag{4.35}$$

where $c = c(n, \gamma) > 0$. Now let

$$Y_i = \frac{1}{M^2 r^n} \int_{B(y, r_i)} |(u - k_i)^+|^2 \, dx$$

and observe that

$$Y_i \ge \frac{(k_{i+1} - k_i)^2}{M^2 r^n} |\{x \in B(y, r_i) : u(x) > k_{i+1}\}| \ge \frac{1}{2^{2i+2} r^n} |A_i|. \tag{4.36}$$

Thus, combining (4.34), (4.35) and (4.36) yields

$$Y_{i+1} \leq cb^i Y_i^{1+\delta}, \quad b = 2^{2+\frac{4}{n}}, \ \delta = 1 + \frac{2}{n}.$$

It is easy to prove (exercise) that there is $\varepsilon_0 = \varepsilon_0(c, b, \delta) > 0$ such that, if $Y_0 \leq \varepsilon_0$, then $Y_i \to 0$ as $i \to \infty$. This means that

$$\sup_{B(y,r/2)} u \leq M$$

if we choose M such that

$$Y_0 = \frac{1}{M^2 r^n} \int_{B(y,r)} |u^+|^2 \, dx = \varepsilon_0.$$

This proves the theorem. $\qquad\qquad\qquad\qquad\qquad\qquad\qquad\qquad\quad$ $\square$

4.4.3 Hölder continuity of functions in $DG(\Omega, \gamma)$

In this subsection, we prove Hölder continuity for functions in the De Giorgi class.

Theorem 4.36. *Let $\gamma > 0$ be a constant. There is an exponent $\alpha = \alpha(n, \gamma)$ with $0 < \alpha \leq 1$ such that, for every $u \in DG(\Omega, \gamma)$, we have $u \in C^{0,\alpha}(\Omega)$. Moreover, there is $\delta = \delta(n, \gamma)$ with $0 < \delta < 1$ such that, for every ball $B(y, r) \subset \Omega$, we have*

$$\mathrm{osc}_{B(y,r/4)} u \leq \delta \, \mathrm{osc}_{B(y,r)} u.$$

Theorem 4.36 follows from the following two lemmas.

Lemma 4.37. *For any $\theta > 0$, there is $s = s(\theta, \gamma, n) \geq 1$ such that the following holds: for every $u \in DG^+(\Omega, \gamma)$ and for every ball $B(y, r) \subset \Omega$, if*

$$|\{x \in B(y, r/2) : u(x) \leq k_0\}| \geq \frac{1}{2} |B(y, r/2)| \qquad (4.37)$$

holds for some $k_0 \in \mathbb{R}$, then we have

$$\left| \left\{ x \in B(y, r/2) : u(x) > M_r - \frac{1}{2^s}(M_r - k_0) \right\} \right| \leq \theta |B(y, r/2)|, \qquad (4.38)$$

where $M_r = \sup_{B(y,r)} u$.

Proof. Fix $B(y, r) \subset \Omega$ and $k_0 \in \mathbb{R}$ such that (4.37) holds. We may assume that $k_0 < M_r$. Otherwise, there is nothing to prove. Since $u \in DG^+(\Omega, \gamma)$, we have, for every k,

$$\int_{B(y,r/2)} |\nabla(u - k)^+|^2 \, dx \leq \frac{4\gamma}{r^2} \int_{B(y,r)} |(u - k)^+|^2 \, dx$$

$$\leq c(n, \gamma) r^{n-2} (M_r - k)^2.$$

For any $k_0 \leq k < l < M_r$, we have, by Hölder's inequality,

$$\left(\int_{A_{k,l}} |\nabla u| \, dx \right)^2 \leq \int_{A_{k,l}} |\nabla u|^2 \, dx \, |A_{k,l}|$$

$$\leq \int_{B(y,r/2)} |\nabla (u-k)^+|^2 \, dx \, |A_{k,l}|,$$

where $A_{k,l} = \{x \in B(y, r/2) \, : \, k < u(x) \leq l\}$. Thus,

$$\left(\int_{A_{k,l}} |\nabla u| \, dx \right)^2 \leq c(M_r - k)^2 r^{n-2} |A_{k,l}|. \tag{4.39}$$

Now note that

$$|\{x \in B(y, r/2) \, : \, u(x) \leq k\}| \geq \frac{1}{2} |B(y, r/2)|$$

for all $k \geq k_0$, due to (4.37). We apply Theorem 4.13 to obtain that

$$(l - k)|A_l|^{1 - \frac{1}{n}} \leq c(n) \int_{A_{k,l}} |\nabla u| \, dx, \tag{4.40}$$

where

$$A_l = \{x \in B(y, r/2) \, : \, u(x) > l\}.$$

It follows from (4.39) and (4.40) that

$$(l - k)^2 |A_l|^{2 - \frac{2}{n}} \leq c(M_r - k)^2 r^{n-2} |A_{k,l}| \tag{4.41}$$

for all $k_0 \leq k < l < M_r$. Now set

$$k_i = M_r - \frac{1}{2^i}(M_r - k_0), \quad i = 0, 1, 2, \ldots.$$

Then (4.41) with $k = k_i$, $l = k_{i+1}$ gives

$$|A_{k_{i+1}}|^{2 - \frac{2}{n}} \leq c r^{n-2} |A_{k_i, k_{i+1}}|.$$

Let $i_0 \in \mathbb{N}$, to be chosen soon. Sum the above inequality from $i = 0$ to $i_0 - 1$. Note that $|A_{k_{i+1}}| \geq |A_{k_{i_0}}|$ for all $0 \leq i \leq i_0 - 1$. Thus we arrive at

$$i_0 |A_{i_0}|^{2 - \frac{2}{n}} \leq c r^{n-2} |A_{k_0}| \leq c r^{2n-2}.$$

By choosing i_0 big enough, we have proved the lemma. $\square$

Lemma 4.38. *There is $\theta = \theta(n, \gamma) > 0$ such that the following holds: for every $u \in DG^+(\Omega, \gamma)$, every ball $B(y, r) \subset \Omega$, and any $k_0 \in \mathbb{R}$, if*

$$|\{x \in B(y, r) : u(x) > k_0\}| \le \theta |B(y, r)|, \tag{4.42}$$

then we have

$$\sup_{B(y, r/2)} u \le \frac{1}{2}k_0 + \frac{1}{2} \sup_{B(y, r)} u. \tag{4.43}$$

Proof. Fix $B(y, r) \subset \Omega$ and fix k_0 such that (4.42) holds for some θ, which will be chosen later. Denote $M_r = \sup_{B(y,r)} u$. We may assume that $k_0 < M_r$. Otherwise, the conclusion of the lemma is trivial. Now we consider a sequence

$$k_i = \frac{k_0}{2} + \frac{M_r}{2} - \frac{1}{2^{i+1}}(M_r - k_0), \quad i = 0, 1, 2, \ldots$$

and a sequence of radii

$$r_i = \frac{r}{2} + \frac{r}{2^{i+1}}, \quad i = 0, 1, 2, \ldots.$$

Since $u \in DG^+(\Omega, \gamma)$, (4.32) with $k = k_i$, $r = r_i$, $r' = r_{i+1}$ gives

$$\int_{A_{k_i, r_{i+1}}} |\nabla u|^2 \, dx \le \frac{\gamma}{(r_i - r_{i+1})^2} \int_{A_{k_i, r_i}} |(u - k_i)^+|^2 \, dx$$

$$\le \frac{c 2^{2i}}{r^2}(M_r - k_0)^2 |A_{k_i, r_i}|, \tag{4.44}$$

where we denote

$$A_{k, \rho} = \{x \in B(y, \rho) : u(x) > k\}.$$

On the other hand, due to (4.42), we have

$$|A_{k_i, r_{i+1}}| \le |A_{k_0, r}| \le \theta |B(y, r)| \le \frac{1}{2}|B(y, r_{i+1})|,$$

if we assume that $\theta \le 1/2^{n+1}$. Then, by Theorem 4.13 with $k = k_i$, $l = k_{i+1}$, we have

$$(k_{i+1} - k_i)|A_{k_{i+1}, r_{i+1}}|^{1 - \frac{1}{n}} \le c(n) \int_{A_{k_i, r_{i+1}}} |\nabla u| \, dx$$

$$\le c(n) \left(\int_{A_{k_i, r_{i+1}}} |\nabla u|^2 \, dx \right)^{\frac{1}{2}} |A_{k_i, r_{i+1}}|^{\frac{1}{2}}. \tag{4.45}$$

Combining (4.44) and (4.45) yields

$$(k_{i+1} - k_i)|A_{k_{i+1}, r_{i+1}}|^{1 - \frac{1}{n}} \le \frac{c 2^i}{r}(M_r - k_0)|A_{k_i, r_i}|^{\frac{1}{2}}|A_{k_i, r_{i+1}}|^{\frac{1}{2}} \le \frac{c 2^i}{r}|A_{k_i, r_i}|,$$

that is,

$$Y_{i+1} \leq cb^i Y_i^{\frac{n}{n-1}}, \quad b = 2^{\frac{2n}{n-1}}, \quad c = c(n, \gamma) > 0, \tag{4.46}$$

where we denote

$$Y_i = \frac{|A_{k_i, r_i}|}{|B(y, r)|}.$$

Now we can iterate (4.46) and prove (exercise) that there is $\theta = \theta(b, c) > 0$ such that $Y_i \to 0$ as $i \to \infty$, provided that $Y_0 \leq \theta$. This proves the lemma. $\quad\square$

Now we are in the position to prove Theorem 4.36.

Proof of Theorem 4.36. Let $u \in DG(\Omega, \gamma)$ and fix a ball $B(y, r) \subset \Omega$. Set

$$M_s = \sup_{B(y,s)} u, \quad m_s = \inf_{B(y,s)} u.$$

Then we let $k_0 = (M_r + m_r)/2$. We may assume that

$$|\{x \in B(y, r/2) : u(x) \leq k_0\}| \geq \frac{1}{2}|B(y, r/2)|.$$

Otherwise, we consider $-u$ instead of u. Let $\theta = \theta(n, \gamma) > 0$ be the number determined as in Lemma 4.38. We apply Lemma 4.37 to obtain that

$$|\{x \in B(y, r/2) : u(x) > M_r - \frac{1}{2^s}(M_r - k_0)\}| \leq \theta|B(y, r/2)|$$

where $s = s(n, \gamma) \geq 1$. Next, we apply Lemma 4.38 to obtain that

$$M_{r/4} \leq \frac{1}{2}\left[M_r - \frac{1}{2^s}(M_r - k_0)\right] + \frac{1}{2}M_{r/2} \leq M_r - \frac{1}{2^{s+1}}(M_r - k_0).$$

Therefore, we have

$$M_{r/4} - m_{r/4} \leq M_r - \frac{1}{2^{s+1}}(M_r - k_0) - m_r = \left(1 - \frac{1}{2^{s+2}}\right)(M_r - m_r),$$

which proves the theorem. $\quad\square$

4.5 Further discussions

4.5.1 Degenerate elliptic equations

In equation (4.1), we assume the following for the coefficients $\mathbb{A}(x)$:

$$\lambda(x)|\xi|^2 \leq \langle \mathbb{A}(x)\xi, \xi \rangle = \sum_{i,j=1}^n a_{ij}(x)\,\xi_i\xi_j \leq \Lambda(x)|\xi|^2 \tag{4.47}$$

for all $\xi = (\xi_1, \ldots, \xi_n) \in \mathbb{R}^n$ and almost every $x \in \Omega$. Here $0 \leq \lambda(x) \leq \Lambda(x) \leq \infty$ are given functions.

In 1995, De Giorgi gave a talk in Lecce, Italy, and discussed the following natural question: Are size assumptions on $\lambda(x)^{-1}$ and $\Lambda(x)$ sufficient to guarantee the continuity of weak solutions? He raised the following conjectures on the continuity of weak solutions of equation (4.1). The first one concerns the singular case in higher dimensions.

Conjecture 4.39 (De Giorgi [2]). Let $n \geq 3$. Suppose that $\mathbb{A}(x)$ satisfies (4.47) with $\lambda(x) = 1$ and with $\Lambda(x)$ such that

$$\int_\Omega \exp(\Lambda(x)) \, dx < \infty. \tag{4.48}$$

Then all weak solutions of equation (4.1) are continuous in Ω.

The second one concerns the degenerate case in higher dimensions.

Conjecture 4.40 (De Giorgi [2]). Let $n \geq 3$. Suppose that $\mathbb{A}(x)$ satisfies (4.47) with $\Lambda(x) = 1$ and with $\lambda(x)$ such that

$$\int_\Omega \exp(\lambda(x)^{-1}) \, dx < \infty. \tag{4.49}$$

Then all weak solutions of equation (4.1) are continuous in Ω.

The third one concerns the case of singular and degenerate equations in higher dimensions.

Conjecture 4.41 (De Giorgi [2]). Let $n \geq 3$. Suppose that $\mathbb{A}(x)$ satisfies (4.47) with $\Lambda(x) = \lambda(x)^{-1}$ such that

$$\int_\Omega \exp(\Lambda(x)^2) \, dx < \infty. \tag{4.50}$$

Then all weak solutions of equation (4.1) are continuous in Ω.

The fourth one concerns the planar case, $n = 2$, which is different from the higher-dimensional cases.

Conjecture 4.42 (De Giorgi [2]). Let $n = 2$. Suppose that $\mathbb{A}(x)$ satisfies (4.47) with $\Lambda(x) = 1$ and with $\lambda(x)$ such that

$$\int_\Omega \exp(\sqrt{\lambda(x)^{-1}}) \, dx < \infty. \tag{4.51}$$

Then all weak solutions of equation (4.1) are continuous in Ω.

Conjectures 4.39, 4.40 and 4.41 are still open. As far as we know, the best known result is due to Trudinger [20], which is far from the conjectures. It seems that one needs new ideas to deal with these challenging problems. Concerning Conjecture 4.42, Onninen and the author [16] recently proved that all weak solutions of equation (4.1) are continuous under the assumption that

$$\int_\Omega \exp(\alpha\sqrt{\lambda(x)^{-1}}) \, dx < \infty$$

for some constant $\alpha > 1$.

Another interesting part of [2] is that De Giorgi also conjectured that his conjectures above are sharp, namely the integrability conditions (4.48), (4.49) and (4.51) are optimal to guarantee the continuity of weak solutions. For example, in Conjecture 4.39, one cannot replace (4.48) by the following weaker one:

$$\int_\Omega \exp(\alpha \Lambda(x)^{1-\delta})\, dx < \infty \qquad (4.52)$$

for some $\delta > 0$ and any $\alpha > 0$. De Giorgi conjectured that one can construct a function $\Lambda(x)$ satisfying the integrability condition (4.52) such that equation (4.1) satisfying (4.47) with $\lambda(x) = 1$ and with this $\Lambda(x)$ has discontinuous weak solutions.

In [2], De Giorgi even gave hints on how to construct counterexamples to show the sharpness of the above conjectures. He made the following precise conjectures. Let $\Omega = \{x = (x_1, \ldots, x_n) \in \mathbb{R}^n : |x| < 1/e\}$ and let A, B be subsets of Ω defined as

$$A = \{x \in \Omega : 2|x_n| > |x|\}, \quad B = \{x \in \Omega : 2|x_n| < |x|\}.$$

The first conjecture would yield the sharpness of Conjecture 4.39.

Conjecture 4.43 (De Giorgi [2]). Let $n \geq 3$. For any $\varepsilon > 0$, define a function τ_1 in Ω as follows:

$$\tau_1(x) = \begin{cases} |\log|x||^{1+\varepsilon} & \text{if } x \in A; \\ 1 & \text{if } x \in B. \end{cases}$$

Then equation (4.1) with $\mathbb{A}(x) = \tau_1(x)I$ has a weak solution discontinuous at the origin.

The second one would yield the sharpness of Conjecture 4.40.

Conjecture 4.44 (De Giorgi [2]). Let $n \geq 3$. For any $\varepsilon > 0$, define a function τ_2 in Ω as follows:

$$\tau_2(x) = \begin{cases} 1 & \text{if } x \in A; \\ |\log|x||^{-(1+\varepsilon)} & \text{if } x \in B. \end{cases}$$

Then equation (4.1) with $\mathbb{A}(x) = \tau_2(x)I$ has a weak solution discontinuous at the origin.

The third one concerns the planar case, and would yield the sharpness of Conjecture 4.42.

Conjecture 4.45 ([De Giorgi [2]). Let $n = 2$. For any $\varepsilon > 0$, define a function τ_3 in Ω as follows:

$$\tau_3(x) = \begin{cases} 1 & \text{if } x \in A; \\ |\log|x||^{-(2+\varepsilon)} & \text{if } x \in B. \end{cases}$$

Then equation (4.1) with $\mathbb{A}(x) = \tau_3(x)I$ has a weak solution discontinuous at the origin.

In [22] it was proved that Conjectures 4.43, 4.44 and 4.45 are true.

Bibliography

[1] De Giorgi, E. *Sulla differenziabilità e l'analiticità delle estremali degli integrali multipli regolari*. Mem. Accad. Sci. Torino Cl. Sci. Fis. Mat. Nat. **3** (1957), 25–43.

[2] De Giorgi, E. *Congetture sulla continuità delle soluzioni di equazioni lineari ellittiche autoaggiunte a coefficienti illimitati*. Unpublished (1995).

[3] Di Benedetto, E. and Trudinger, N. S. *Harnack inequalities for quasiminima of variational integrals*. Ann. Inst. H. Poincaré Anal. Non Linéaire **1** (1984), no. 4, 295–308.

[4] Evans, L. C. *Partial Differential Equations*. Graduate Studies in Math., vol. 19, Amer. Math. Soc., Providence, RI, 1998.

[5] Fabes, E. B., Kenig, C. E. and Serapioni, R. *The local regularity of solutions of degenerate elliptic equations*. Comm. Partial Differential Equations **7** (1982), 77–116.

[6] Fabes, E. B. and Stroock, D. W. *A new proof of Moser's parabolic Harnack inequality using the old ideas of Nash*. Arch. Rational Mech. Anal. **96** (1986), no. 4, 327–338.

[7] Gilbarg, D. and Trudinger, N. S. *Elliptic Partial Differential Equations of Second Order*. 2nd ed., Springer-Verlag, New York, 1983.

[8] Iwaniec, T. and Sbordone, C. *Weak minima of variational integrals*. J. Reine Angew. Math. **454** (1994), 143–161.

[9] Ladyzhenskaya, O. A. and Ural'tseva, N. N. *Linear and Quasilinear Elliptic Equations*. Academic Press, New York, 1968.

[10] Meyers, N. G. *An L^p-estimate for the gradient of solutions of second order elliptic divergence equations*. Ann. Sc. Norm. Super. Pisa **17** (1963), 189–206.

[11] Morrey, C. B. *On the solutions of quasi-linear elliptic partial differential equations*. Trans. Amer. Math. Soc. **43** (1938), 126–166.

[12] Morrey, C. B. *Multiple integral problems in the calculus of variations and related topics*. Univ. California Publ. Math. (N.S.) **1** (1943), 1–130.

[13] Moser, J. *A new proof of De Giorgi's theorem concerning the regularity problem for elliptic differential equations*. Comm. Pure Appl. Math. **13** (1960), 457–468.

[14] Moser, J. *On Harnack's theorem for elliptic differential equations*. Comm. Pure Appl. Math. **14** (1961), 577–591.

[15] Nash, J. *Continuity of solutions of elliptic and parabolic equations*. Amer. J. Math. **80** (1958), 931–954.

[16] Onninen, J. and Zhong, X. *Continuity of solutions of linear, degenerate elliptic equations*. Ann. Sc. Norm. Super. Pisa Cl. Sci. (5) **6** (2007), 103–116.

[17] Piccinini, L. C. and Spagnolo, S. *On the Hölder continuity of solutions of second order elliptic equations in two variables.* Ann. Sc. Norm. Super. Pisa (3) **26** (1972), 391–402.

[18] Serrin, J. *On the strong maximum principle for quasilinear second order differential inequalities.* J. Funct. Anal. **5** (1970), 184–193.

[19] Trudinger, N. S. *On Harnack type inequalities and their application to quasilinear, elliptic equations.* Comm. Pure Appl. Math. **20** (1967), 721–747.

[20] Trudinger, N. S. *On the regularity of generalized solutions of linear, nonuniformly elliptic equations.* Arch. Rational Mech. Anal. **42** (1971), 51–62.

[21] Widman, K. O. *On the Hölder continuity of solutions of elliptic partial differential equations in two variables with coefficients in L^∞.* Comm. Pure Appl. Math. **22** (1969), 669–682.

[22] Zhong, X. *Discontinuous solutions of linear, degenerate elliptic equations.* J. Math. Pures Appl. (9) **90** (2008), 31–41.

GPSR Compliance

The European Union's (EU) General Product Safety Regulation (GPSR)
is a set of rules that requires consumer products to be safe and our
obligations to ensure this.

If you have any concerns about our products, you can contact us on
ProductSafety@springernature.com

In case Publisher is established outside the EU, the EU authorized
representative is:

Springer Nature Customer Service Center GmbH
Europaplatz 3
69115 Heidelberg, Germany

Batch number: 09473855

Printed by Printforce, the Netherlands